WIND LOADS FOR PETROCHEMICAL AND OTHER INDUSTRIAL FACILITIES

PREPARED BY
Task Committee on Wind-Induced Forces of the
Petrochemical Committee of the
Energy Division of the
American Society of Civil Engineers

1801 ALEXANDER BELL DRIVE
RESTON, VIRGINIA 20191–4400

Cataloging-in-Publication Data on file with Library of Congress

American Society of Civil Engineers
1801 Alexander Bell Drive
Reston, Virginia, 20191-4400

www.asce.org/pubs

Any statements expressed in these materials are those of the individual authors and do not necessarily represent the views of ASCE, which takes no responsibility for any statement made herein. No reference made in this publication to any specific method, product, process, or service constitutes or implies an endorsement, recommendation, or warranty thereof by ASCE. The materials are for general information only and do not represent a standard of ASCE, nor are they intended as a reference in purchase specifications, contracts, regulations, statutes, or any other legal document. ASCE makes no representation or warranty of any kind, whether express or implied, concerning the accuracy, completeness, suitability, or utility of any information, apparatus, product, or process discussed in this publication, and assumes no liability therefore. This information should not be used without first securing competent advice with respect to its suitability for any general or specific application. Anyone utilizing this information assumes all liability arising from such use, including but not limited to infringement of any patent or patents.

ASCE and American Society of Civil Engineers—Registered in U.S. Patent and Trademark Office.

Photocopies and permissions. Permission to photocopy or reproduce material from ASCE publications can be obtained by sending an e-mail to permissions@asce.org or by locating a title in ASCE's online database (http://cedb.asce.org) and using the "Permission to Reuse" link. *Bulk reprints.* Information regarding reprints of 100 or more copies is available at http://www.asce.org/reprints.

Copyright © 2011 by the American Society of Civil Engineers.
All Rights Reserved.
ISBN 13: 978-0-7844-1180-3
Manufactured in the United States of America.

16 15 14 13 12 11   1 2 3 4 5 6 7

ASCE Petrochemical Energy Committee

This publication is one of five state-of-the-practice engineering reports produced, to date, by the ASCE Petrochemical Energy Committee. These engineering reports are intended to be a summary of current engineering knowledge and design practice, and present guidelines for the design of petrochemical facilities. They represent a consensus opinion of task committee members active in their development. These five ASCE engineering reports are:
1) *Design of Anchor Bolts in Petrochemical Facilities*
2) *Design of Blast Resistant Buildings in Petrochemical Facilities*
3) *Design of Secondary Containment in Petrochemical Facilities*
4) *Guidelines for Seismic Evaluation and Design of Petrochemical Facilities*
5) *Wind Loads for Petrochemical and Other Industrial Facilities*

The ASCE Petrochemical Energy Committee was organized by A. K. Gupta in 1991 and initially chaired by Curley Turner. Under their leadership the five task committees were formed. More recently, the Committee has been chaired by Joseph A. Bohinsky and Frank J. Hsiu. The five reports were initially published in 1997.

Buildings codes and standards have changed significantly since the publication of these five reports, specifically in the calculation of wind and seismic loads and analysis procedures for anchorage design. Additionally, new research in these areas and in blast resistant design has provided opportunities for improvement of the recommended guidelines. The ASCE has determined the need to update four of the original reports and publish new editions, based on the latest research and for consistency with current building codes and standards.

The ASCE Petrochemical Energy Committee was reorganized by Magdy H. Hanna in 2005 and the following four task committees were formed to update their respective reports:
- Task Committee on Anchor Bolt Design for Petrochemical Facilities
- Task Committee on Blast Design for Petrochemical Facilities
- Task Committee on Seismic Evaluation and Design for Petrochemical Facilities
- Task Committee for Wind Load Design for Petrochemical Facilities

Current ASCE Petrochemical Energy Committee

Magdy H. Hanna	Jacobs Engineering - Chairman
William Bounds	Fluor Corporation
John Falcon	Jacobs Engineering
James R. (Bob) Bailey	Exponent, Inc.
J. G. (Greg) Soules	CB&I

The ASCE Task Committee on Wind-Induced Forces

This report is intended to be a state-of-the-practice set of guidelines. It is based on reviews of current practice, internal company standards, published documents, and the work of related organizations. The report includes a list of references that provides additional information. The reference list emphasizes readily available commercial publications and government reports.

This report was prepared to provide guidance for determination of wind induced forces on structures found in petrochemical and other industrial facilities. It should be of interest to engineers familiar with design of industrial type structures and the application of ASCE 7, "Minimum Design Loads for Buildings and other Structures," to these types of structures.

In helping create a consensus set of guidelines, a number of individuals provided valuable assistance and review. Reviewers included John Geigel (ExxonMobil), Drew Troyer (ConocoPhillips), and Eric Wey (Fluor Corporation). The committee is appreciative of the efforts of these reviewers.

The task committee would also like to acknowledge the numerous contributions made to this task committee and other technical committees over the years by both Michael Bergeron (SNC Lavalin – GDS Engineers) and Mike Chen (Fluor Corporation). Both Michael and Mike passed away during the preparation of this report update and will be sorely missed by the committee and the broader engineering community.

Finally, the committee would also like to thank Judy Falcon (Exponent, Inc.) who patiently and diligently edited the manuscript and put up with all of our changes.

The ASCE Task Committee on Wind-Induced Forces

James R. (Bob) Bailey Ph.D., P.E., F. ASCE	Richard T. Gilbert P.E.	Paul B. Summers P.E., S.E.
Exponent, Inc.	ExxonMobil Research & Engineering Company	MMI Engineering
Chairman	Co-Chairman	Secretary

Samuel D. Amoroso, Ph.D., P.E.	Engensus Engineering & Consulting
K.C. Fong, P.E.	URS Corporation
Javier Garza, P.E.	Shell
Madgy H. Hanna (Past Co-Chairman)	Jacobs Engineering
Don Harnly, P.E.	Jacobs Engineering
Kirby Hebert	Louisiana State University
Marc L. Levitan (Past Chairman)	Louisiana State University
Guzhao Li, Ph.D., P.E., S.E.	MMI Engineering
Gerald W. Mayes, P.E.	The Shaw Group
Rajendra Prasad, P.E., PMP	Mustang Engineering, L.P.
Norman Rennalls, P.E.	Shell
Amy Styslinger, P.E.	ExxonMobil Upstream Research Company
Walter A. Waller, P.E.	Bechtel Corp
James H. Wissehr, P.E., S.E.	Jacobs Engineering
Silky S.K. Wong, P.E., S.E.	Fluor Corporation
Randall L. Wright, P.E., S.E.	Mustang Engineering, L.P.
Gregory B. Young, P.E.	ConocoPhillips

Table of Contents

Chapter 1: Introduction 1

    1.1   Background 1
    1.2   State of the Practice 2
    1.3   Purpose of Report 2

Chapter 2: Background 3

    2.1   Introduction 3
    2.2   Key Wind Engineering Concepts 3
    2.3   Aerodynamics of Open Frame Structures 9
    2.4   Aerodynamics of Partially Clad Structures 10
    2.5   Aerodynamics of Vertical Vessels 11
    2.6   Other Wind Loading Codes, Standards, and Guides 14
    2.7   Research Progress and Future Needs 15

Chapter 3: Review of Existing Design Practices 18

    3.1   Introduction 18
    3.2   Survey of Existing Practices and Impacts of First Edition Guidelines 18
    3.3   Existing Design Practices for Structures New to Second Edition 21

Chapter 4: Recommended Guidelines Part I: Design Considerations and Methods 31

    4.0   Introduction 31
    4.1   Historical Performance 31
    4.2   Wind Tunnel Testing 43
    4.3   CFD Commentary 45
    4.4   Load Combinations 46
    4.5   Special Considerations for LNG Facilities 55
    4.6   Evaluation of Wind Loads on Existing Structures 58
    4.7   Wind Load Analysis Uncertainty 64

Chapter 5: Recommended Guidelines Part II: Analytical Determination of Wind Loads 66

    5.0   General 66
    5.1   Pipe Racks 67
    5.2   Open Frame Structures 69
    5.3   Partially Clad Structures 83

5.4	Pressure Vessels	85
5.5	Cooling Towers	94
5.6	Air Cooled Heat Exchanger (Air Coolers or Fin Fans)	95

Appendix 5A	Alternate Method for Determining C_f and Load Combinations for Open Frame Structures	97
Appendix 5B	High-Solidity Open Frame Structures	106

Chapter 6: Examples 110

6.0	Introduction	110
6.1	Pipe Rack and Pipe Bridge Example	111
6.2	Open Frame Examples	123
6.3	Partially Clad Structure Example	137
6.4	Pressure Vessels Example	140
6.5	Cooling Tower Example	154

References 160

Index 167

CHAPTER 1
INTRODUCTION

The focus of this report is on the procedures for determining the design wind loads for non-building structures in petrochemical and other industrial facilities. The report is structured around the following generic types of structures usually found in these facilities. Examples are also provided for some of these structures:

   a. Pipe support structures (pipe racks, pipe bridges)
   b. Open and partially clad frame structures
   c. Vessels (vertical, horizontal and spherical)
   d. Cooling towers
   e. Air coolers (air cooled heat exchangers, also known as fin fans)
   f. Tanks
   g. Steel stacks

1.1 Background

The basis and procedures for determining design wind loads for enclosed structures and other conventional structures are well documented in the engineering literature. These design basis and procedures have been adopted by ASCE and prescribed in ASCE/SEI 7-05[1] (herein referred to as *ASCE 7*) and its predecessor documents. Other organizations have incorporated the major provisions of *ASCE 7* into building codes. The International Building Code (IBC) states that wind loads should be calculated in accordance with *ASCE 7*, and the IBC has been adopted throughout the United States. *ASCE 7* provides three methods for calculating design wind loads on the main wind force resisting system (MWFRS) and on components and cladding:

   1. Simplified procedure
   2. Analytical procedure
   3. Wind tunnel procedure

The simplified procedure (Method 1 in *ASCE 7*) was introduced to simplify the analysis of typical regular-shaped building structures. Its use is restricted to relatively rigid, low-rise, enclosed structures. The analytical procedure (Method 2 in *ASCE 7*) is permitted for structures of any height that do not have unusual geometric irregularities or unusual response characteristics. The wind tunnel procedure is required for complex structures that cannot be evaluated using Method 1 or 2.

The Scope statement for *ASCE 7* indicates that the standard provides minimum load requirements for the design of buildings and other structures that are subject to building codes. ASCE 7 also addresses enclosed structures, trussed towers, and simple cylinders commonly found in petrochemical facilities. However, to address important non-building structures in petrochemical and other industrial facilities, this report enhances ASCE 7 provisions for open frame structures, structures with

---

[1] At the time of publication of this report, ASCE/SEI 7-10 had been released.

interconnecting piping, partially clad structures, vessels with attached piping and platforms, cooling towers, and air coolers. Design wind loads on non-building structures are typically calculated using the force equation from *ASCE 7*:

$$F = q_z \, G \, C_f \, A_f \qquad \qquad (ASCE\ 7\ \text{Eq. 6-28})$$

In this equation q_z is the velocity pressure component, G is the gust effect factor, C_f is the force/shape/drag/shielding component, and A_f is the area for which the force is calculated that is usually the projected area normal to the wind. The velocity pressure, q_z, is calculated using *ASCE 7* based on several factors, such as the importance of the structure, the surrounding terrain (exposure category), and the basic wind speed, among others.

The selection of basic wind speed, importance factor, exposure category, gust effect factor, and other factors is described in *ASCE 7* and, therefore, is not discussed in detail herein. This report also expands upon ASCE 7 coverage of force coefficients, tributary areas, and shielding for industrial type structures and equipment, which must be carefully defined to assure behavior under wind forces are accounted for. These wind load components are discussed in this report and recommendations for selecting values are made. Since this report is intended to supplement *ASCE 7*, the designer is referred to that document when it provides the appropriate information. The nomenclature used in the recommendations of this report mirrors those found in *ASCE 7*.

1.2 State of the Practice

This report reflects various company practices, available research and committee consensus for the wind load design of petrochemical and other industrial facilities. The committee performed a survey (see Section 3.2) and the results indicate that more than half of the companies surveyed have design practices that reference the first edition of this report released in 1997. These survey results are evidence that there has been increased uniformity from a decade ago in determining wind loads on petrochemical structures.

1.3 Purpose of Report

It is the intent of this committee that the publication of this report will continue to progress a more uniform application of practices for the computation of design wind loads for petrochemical and industrial facilities. In order to facilitate this goal, a set of recommended guidelines is presented as part of this report.

CHAPTER 2
BACKGROUND

2.1 Introduction

This chapter provides background information that is intended to assist the user of this document in the interpretation and application of the material that is presented in subsequent chapters. The sections of this chapter include discussions of fundamental wind engineering concepts and the general aerodynamic characteristics of structure types common to petrochemical and other industrial facilities. The chapter concludes with a presentation of other sources of guidance that may be helpful to designers or analysts, a brief summary of research progress since the last edition of this guide, and research needs for the future.

2.2 Key Wind Engineering Concepts

Wind Load Formulation

The wind load on a structure is a function of many different variables. The purpose of this section is to review the basic wind load formulation in *ASCE 7* as it pertains to petrochemical structures and briefly discuss some considerations relevant to each of the variables. Equation 6-28 in *ASCE 7* is used to calculate the wind force on other, non-building structures such as petrochemical structures. This equation is shown below. The four variables in this equation are the velocity pressure, q_z, the gust effect factor, G, the aerodynamic force coefficient, C_f, and the reference area, A_f.

$$F = q_z \cdot G \cdot C_f \cdot A_f \qquad\qquad (ASCE\ 7\ \text{Eq. 6-28})$$

The velocity pressure, q_z, as defined by Equation 6-15 in *ASCE 7* and shown below in US customary units (for the metric equivalent of this equation see *ASCE 7*) is a modification of Bernoulli's formula, which relates the fluid dynamic pressure to the fluid density, and the square of the flow velocity (or in this case, the basic wind speed). The constant at the beginning of the velocity pressure equation in *ASCE 7* includes the density of air and a factor for the conversion of units such that when the wind speed is specified in units of miles per hour, the resulting velocity pressure is in units of pounds per square foot. The remaining variables in the velocity pressure equation are dimensionless.

$$q_z = 0.00256 \cdot K_z \cdot K_{zt} \cdot K_d \cdot V^2 \cdot I \qquad\qquad (ASCE\ 7\ \text{Eq. 6-15})$$

Of all of the variables in this equation, the specification of the basic wind speed, V, has the greatest influence on the velocity pressure, q_z, and the resulting wind load, F. Basic wind speeds are specified by *ASCE 7* as three-second gust wind speeds at standard meteorological heights of 33 feet (10 m) in open terrain with nominal return periods (or mean recurrence intervals) of 50 years. The inverse of the mean recurrence interval (MRI) for a given wind speed in years is the probability that such a wind speed will occur in any given year. Since some geographic locations are more susceptible to severe winds than others due to variations in local climatology, the wind speed associated with a particular MRI generally will also vary with geographic location.

ASCE 7 Figure 6-1 is a map of the United States with contours of basic wind speed (isotachs) and political boundaries resolved to the state and county levels. This map facilitates easy selection of basic wind speeds for use in *ASCE 7* Equation 6-15. Locations away from the eastern seaboard and the coast of the Gulf of Mexico in the United States have been found to be approximately meteorologically homogeneous as far as extreme wind speeds are concerned. For these locations, the 50-year gust wind speed has been synthesized statistically using meteorological observations from hundreds of measurement stations over the course of several decades. Due to the general lack of reliable meteorological measurements in the hurricane-prone regions of the Atlantic and Gulf coasts of the United States, the basic wind speeds for these locations have been estimated through Monte Carlo simulation using mathematical models of hurricane wind fields and statistical data gathered over the past 150 or so years. Special wind regions that have unique meteorological influences are identified on the map in *ASCE 7* Figure 6-1. Specification of the basic wind speed for special wind regions requires consultation with the local building official.

Wind speeds typically increase with height above the ground surface. This variation occurs because of the influences of friction and viscosity on the wind field. The rate at which the wind speed increases with height is governed primarily by the roughness of the terrain over which it is flowing. This roughness may be in the form of elements of the built environment, vegetation, or water waves on large bodies of water located in the upwind fetch. The combined effects of structure height and surface roughness are incorporated into the estimate of the velocity pressure, q_z, through the exposure factor, K_z. This factor is specified in *ASCE 7* for three different terrain exposures and for a variety of heights. The three terrain exposures correspond to urban/suburban or forested terrain (Exposure B), open terrain or large bodies of water in hurricane-prone regions (Exposure C), and very flat terrain or shoreline locations in non-hurricane regions (Exposure D). The value of the exposure factor increases with increasing height above the ground and decreasing surface roughness. Since the basic wind speed, V, is defined at the standard height of 33 feet (10 m) in open terrain, the value of K_z for Exposure C at a height of 33 feet is 1.0.

Topographic effects can influence the wind loading of structures by disturbing the flow field near features such as hills, ridges, and escarpments, resulting in local deviations from the basic wind speed. In general, these changes may increase or decrease the local wind speed, but the *ASCE 7* wind load formulation conservatively only considers cases in which topographic features tend to increase local wind velocities, and thus wind loads. These effects are incorporated into the wind load formulation through an adjustment of the velocity pressure by the topographic effect factor, K_{zt}. *ASCE 7* provides methods for calculating the value of this factor for some simple special cases. The engineer should consider whether or not these special cases adequately represent the conditions for location of interest. For more complicated topography, the engineer may need to exercise careful engineering judgment, seek expert consultation, or resort to either physical or numerical modeling to understand the topographic influences on the wind loads.

The directionality factor, K_d, accounts for the lower probability that the most vulnerable orientation for a structure will correspond unfavorably with the direction of maximum wind speed during a design event. As such, the value of K_d is always less than 1.0, but varies with structure type. *ASCE 7* provides tabulated values of the directionality factor for a variety of structural forms. It is important to note that the values for K_d are defined only for use with the load combinations listed in Section 2.3 and 2.4 of *ASCE 7* when designing a structure. It is the opinion of this committee that the load combinations and examples presented in this document use the appropriate *ASCE 7* values for K_d. The use of the directionality factor may also be unwarranted when using *ASCE 7* methods in an analysis of the performance of an existing structure for which the wind direction is specified or known (e.g. an analysis in a forensic investigation).

The gust effect factor, G, simultaneously incorporates two different processes through which the turbulence in the atmospheric wind can affect wind loads for a structure. The first of these processes is related to the size of a structure relative to significant scales that appear in the turbulence. By definition, a gust of wind is short-lived. Relating wind speed and length of time required for passage of a gust gives some indication of the relative dimensions of the gust. Of course, turbulence, or gustiness, is a chaotic process, and these characteristics can be defined only in a statistical sense. If a structure is large relative to the significant scales of the wind gusts, then the building will often not be completely enveloped by a single gust of wind. As such, the maximum pressures that occur on different building surfaces will not be strictly correlated in time, and the total resulting wind load will be lower than what would be expected by considering the gust wind speed to apply to all the building surfaces simultaneously. *ASCE 7* provides a method for calculating how the relationship between building dimensions and the turbulence characteristics of the wind field influence the gust effect factor.

In addition to the spatial considerations related to wind gusts, the relative flexibility of the structure combined with the temporal fluctuations in the flow velocity may cause certain structures to be dynamically excited in the along-wind direction. If the fundamental frequency of a structure is low enough to enter the range of significant frequency content associated with the wind turbulence, the structure can experience some load amplification due to resonance. *ASCE 7* also provides a method for calculating how the relationship between building fundamental frequency and the turbulence characteristics of the wind field influence the gust effect factor. These dynamic effects can be neglected if a structure is sufficiently rigid. *ASCE 7* defines a rigid structure as one with a fundamental frequency greater than 1 Hz. The structures that may be susceptible to along-wind dynamic excitation in petrochemical facilities are tall, slender structures such as vertical vessels, flare stacks and structures with heavy equipment located or supported near the top of the structure.

The two remaining variables in the wind load formulation are the force coefficient, C_f, and the projected area, A_f. The force coefficient is an empirical factor that typically is determined through experiment for the structure or shape of interest. The determination of values for these variables as they relate to petrochemical structures is the focus of Chapter 5 of this report.

In addition to dependence on structure shape, the force coefficient, C_f, also depends on a non-dimensional flow parameter called the Reynolds number. The Reynolds number (Re) is a ratio of the inertial forces in a flow field to the viscous forces in the flow field, and is defined as follows:

$$Re = \frac{U \cdot D}{v}$$ (Eq. 2-1)

Where U is the flow velocity, D is a characteristic dimension for the structure or element, and v is the kinematic viscosity. In the range of Re that is typical of full-scale flows for structural applications, many structural forms will not exhibit much sensitivity of the force coefficient to Re. This phenomenon is particularly true for rectangular bodies or structures with sharp geometries. Structures with curved surfaces generally show more sensitivity of the force coefficient to Re. For example, the force coefficient for a cylindrical section comprising a handrail component may be substantially larger than the force coefficient for the body of a vertical vessel, which is also cylindrical in form. For the same wind environment, the difference in diameters may be two orders of magnitude, resulting in a proportional difference in Re for the flow near each of these elements. The flow fields and associated force coefficients for cylindrical bodies exhibit dramatic variability for a particular range of Re (10^5 to 10^6). As an example, the force coefficient data provided in Figure 6-21 of *ASCE 7* accounts for these effects for structures such as round chimneys and tanks.

Wind Speed Interpretations

The design wind speeds used in the wind load formulation in *ASCE 7* are specified as "three second gust" wind speeds. This nomenclature refers to the time over which wind speed measurements are averaged. For example, if an anemometer (an instrument for measuring wind speed) records data at a frequency of 10 Hz (i.e. 10 readings per second), then the three-second gust wind speed would be the arithmetic mean of 30 consecutive individual measurements from that device. For fluctuating wind speed signals typical of atmospheric conditions, individual measurements would be higher or lower than the resulting mean value over the averaging interval. Furthermore, it will always be possible to find a subinterval (say one second in this example) that has a mean wind speed higher than that of the interval itself.

This concept extends to other averaging times as well. "Sustained" wind speeds are conventionally defined on the basis of a one-minute averaging time. For any given sustained wind speed measurement, there will be a higher maximum three second gust wind speed occurring in the 60 second interval corresponding to the measurement. Gust factors (not to be confused with gust *effect* factors described previously) are used to convert wind speeds based on one averaging time to equivalent wind speeds corresponding to another averaging time. When converting wind speeds from longer averaging times to shorter averaging times, the gust factors will always be greater than 1.0. The commentary of *ASCE 7* includes a figure which can be used to determine gust factors for converting between equivalent wind speeds for averaging times ranging from one second to one hour. As an example, the factor for converting a one-minute "sustained" wind speed to an equivalent three-second gust wind speed is approximately 1.22 according to the figure in the *ASCE 7* commentary.

A wide variety of wind speed averaging times are represented in engineering standards and through meteorological reporting. *ASCE 7* specified design wind speeds on the basis of the "fastest mile of wind" prior to the 1995 edition of the standard. The fastest mile specification actually has a variable averaging time, which is shown below.

$$t = \frac{3600}{V_{fm}}$$ (Eq. 2-2)

where t is the averaging time in seconds, and V_{fm} is the fastest mile wind speed in miles per hour. The averaging time is equal to 60 seconds for wind speeds of 60 miles per hour, but increases for lower wind speeds and decreases for higher wind speeds. Other international engineering standards have specified design wind speeds on the basis of a mean hourly value or a 10-minute mean. When reviewing older designs or when analyzing structures at international facilities, it is important to recognize that these variations may be present.

Meteorological measurements located at U.S. airports have historically reported gust and sustained wind speeds based on five second and two minute averaging times, respectively. As technology has changed mechanical cup anemometers are being replaced with sonic anemometers at these locations, and the reference averaging times for the measurements are being changed to three seconds and one minute, respectively. Hurricane intensity is conventionally defined on the basis of the fastest one-minute mean wind speed occurring anywhere in the storm (at a reference height of 10 meters or 33 feet above open water). The commentary to *ASCE 7* contains a table providing conversions of the sustained wind speeds for each of the Saffir-Simpson hurricane categories to their equivalent three-second gust values.

Cross-Wind Forces

The *ASCE 7* wind load formulation for other, non-building structures is applicable to loadings occurring along the nominal direction of the wind. *ASCE 7* contains no provisions for estimating wind loads that develop in a direction perpendicular to the wind direction for these types of structures. There are some common instances in which such forces may be significant. If a structure is asymmetric in plan with respect to the wind direction, then it is likely that wind loads will be generated perpendicular to the flow. This scenario may exist for a single structure, or for a structure in a group that is oriented asymmetrically with respect to the wind direction. The aerodynamic action responsible for the development of these types of cross-wind forces is similar to development of lift on the surface of an airplane wing. The unequal velocity and pressure distributions that develop on asymmetric bodies lead to net forces acting perpendicular to the flow direction. These across-wind forces act simultaneously with the along-wind drag forces.

Another form of cross-wind aerodynamic loading can be generated in an entirely different manner. Long, slender structures in relatively non-turbulent wind fields can display wake patterns with regular, alternating vortices. These vortices, or "eddies" in the flow, have rotational axes parallel to the long axis of the structure. The formation of a vortex from one side of the structure results in a flow field in the neighborhood of the structure that is very asymmetric, even though the shape of the structure may not be asymmetric. As discussed in the previous paragraph, the asymmetric flow field will cause a net force on the structure in the across-wind direction. This condition is temporary, however. The development of a complementary vortex on the opposite side of the structure will displace the former vortex, and cause the net cross-wind force to change directions. Cross-wind forces may cancel each other when considering the time averaged effect, even though the alternating forces are significant. The frequency of the vortex shedding pattern is governed by dimension of the cross section, shape of the cross section, and wind speed. The Strouhal number is a non-dimensional parameter that represents the interaction of these effects:

$$St = \frac{n \cdot D}{U} \qquad \text{(Eq. 2-3)}$$

where St is the Strouhal number, n is the frequency of vortex shedding, D, is the cross sectional dimension, and U is the wind speed. If the vortex shedding frequency is near one of the natural frequencies of the structure, an unfavorable amplification of aerodynamic loading can occur. Steel stacks are sometimes susceptible to this action, and for this reason many are outfitted with helical "strakes" in the upper portion of the structure. Helical strakes disrupt formation and longitudinal coherence of the alternating vortices, thus mitigating the dynamic cross-wind loading. The presence of platforms on vertical vessels and towers also generally prevents the formation of these vortices. These features increase the projected area of the structure, and can change the flow pattern such that along-wind forces are increased, but many times this is a worthwhile design concession if resonant oscillations can be avoided.

2.3 Aerodynamics of Open Frame Structures

Open frame structures are common in petrochemical facilities. These structures are characterized by regular bays of open structural framing, the presence of equipment housed within the envelope of the structure, the presence of piping, and other appurtenances such as stairs and handrails. The estimation of wind loads for these structures is complicated by the interactions of the wind with these features.

The wind loads for the frame itself are mainly a function of the solidity of the frame (how much solid area is projected toward the wind), the spacing of the frames, and the total number of frames in the along-wind direction. Upwind frames shield downwind frames to a large degree. Therefore it is overly conservative to calculate the wind load for one frame and multiply this force by the number of frames. It is important to note that the maximum wind force for an open frame structure occurs when the wind approaches the structure at a skewed angle to the axis of the structure (see Figure 5.2). When the wind is oriented along the structure's axis, leeward frames are more directly shielded by their upwind counterparts. When the wind is skewed to the axis of the structure, leeward columns are more exposed to the wind. These downstream members still do not receive the same wind load as the upwind frame since some of the momentum in the flow has been redirected by the first frame line.

The equipment elements housed within an open frame structure contribute to the overall wind load on a structure, and the individual supports for these elements must also be designed to resist wind forces. These elements are typically cylindrical shapes (such as vessels or exchangers). The wind force coefficients for these elements are well-known. However, the equipment will experience shielding from the structural frame, thus reducing the wind load from that of an isolated element. In turn, the equipment shields downwind components of the frame, further reducing the forces from what would otherwise be estimated.

Chapter 5 of this document provides recommended methods for estimating the wind forces for open frame structures and the equipment housed within. There have been some updates and an addition to the methodology that was given in the previous version of this document:

1. The contribution to wind loading of diagonal bracing that is located in a plane parallel to the nominal wind direction is recognized.

2. The wind load-reducing effects of solid floors have been confirmed through research, and the guide contains provisions for taking advantage of this effect.

3. Recent research has uncovered a more accurate (and less conservative) method for estimating the shielding of equipment within open frame structures.

4. An alternative analysis technique for estimating the wind loads for highly-solid open frame structures is presented. This method may be particularly useful for estimating envelope wind loads for structure with equipment layouts that have not yet been determined or are expected to change during the life of the structure.

Chapter 6 includes worked examples for open frame structures demonstrating the application of these updated provisions.

2.4 Aerodynamics of Partially Clad Structures

Structures with partial cladding are sometimes found in petrochemical or process industry facilities. Since process structures are not usually intended for human occupancy, the placement of cladding may not be complete and may be placed to protect only certain equipment, process areas, or individual substructures (for example, stairwells). This section discusses some general aerodynamic characteristics of structures with partial cladding. The material in this section is based on research by Amoroso, et al. (2010). This research considered rectangular structures with no walls clad, some walls clad (in various configurations), and all walls clad. It should be noted that the scenarios represented in this research did not consider all the cases of partial cladding that may be encountered in a petrochemical facility. For example, compressor shelters, which typically consist of steel portal frame buildings with cladding only on the roof and upper portions of the walls, are not covered in this section.

Clearly, the wind effects for the limiting cases of no cladding and full cladding are covered by considering the structure as an open-frame structure or a fully-clad building, respectively. The previous section of this document discusses the general aerodynamics of open-frame structures, and Chapter 5 contains provisions for estimating wind forces for such structures. *ASCE 7* includes pressure coefficients for fully clad buildings.

For rectangular buildings with one, two, or three walls clad, the development of forces depends strongly on the wind direction. For the case of one wall clad, the force coefficients are similar to fully clad buildings when the wind direction is nominally perpendicular to the clad side. When the wind direction is nominally parallel to the clad face, the force coefficients are similar to those of an open frame structure.

Two scenarios are possible for the case of two walls clad: (1) two parallel walls are clad, and (2) two adjacent walls are clad. An important observation from the research was that both of these configurations cause significantly higher wind forces than the

WIND LOADS FOR PETROCHEMICAL AND OTHER INDUSTRIAL FACILITIES

fully clad case. Furthermore, the range of wind directions for which high force coefficients exist extends beyond what is typical of enclosed buildings. For this reason, a designer should consider load cases in which high wind forces are acting simultaneously along both structure axes.

When two parallel walls are clad, the highest force acts in the direction normal to the clad walls. However, the wind direction is from an angle substantially rotated from the orientation of the force. As the wind approaches the structure, the flow is redirected along a path between the parallel, clad walls. The flow is forced to change direction by the presence of the walls, and this change in momentum in the fluid flow is accompanied by a corresponding reaction in the structure. The walls function like vanes in a duct.

When two perpendicular walls are clad, the aerodynamics are quite complex. As one would expect, the maximum forces occur when the clad walls are "cupped" into the wind. However, the maximum force for a given structural axis occurs when the open face is generally oriented to the wind. When the clad faces are not "cupped" into the wind, the development of forces is similar to that on an enclosed building.

For the case of three walls clad, the wind forces are slightly higher than that of an enclosed building when the unclad wall is positioned on the windward face of the structure. Furthermore, high force coefficients persist for a greater range of wind angles than for the fully clad case. When the unclad wall is on the leeward side of the building, the force coefficients are similar to the fully clad case.

Provisions have been included in Chapter 5 for estimating wind forces for partially clad structures. A worked example is also included in Chapter 6 illustrating the application of the provisions. These provisions are for the estimation of overall forces to be resisted by the Main Wind Force Resisting System. Component and Cladding pressures are not provided. Given the high overall force coefficients that can develop on partially clad structures, a designer should exercise caution in using the Component and Cladding Pressures given in *ASCE 7* for application on partially clad structures.

2.5 Aerodynamics of Vertical Vessels

Cylinder Aerodynamics

Vertical vessels are essentially large circular cylinders, and therefore cylinder aerodynamics forms the basis for estimating the wind loads on these types of structures. Force coefficients for circular cylinders in cross flow are well-documented in the fluid mechanics literature. *ASCE 7* provides force coefficient values for circular cylinders in Figure 6-21. As reflected in that figure, the force coefficient for circular cylinders depends on three primary variables: (1) the Reynolds number, (2) the surface roughness of the cylinder, and (3) the aspect ratio, which is the ratio of the cylinder length to the cylinder diameter.

Reynolds Number

The Reynolds number, Re, is the ratio of inertial forces to the viscous forces in a flow and was defined in section 2.2. The aerodynamic force coefficient for circular cylinders is famously sensitive to Re. The reason for this sensitivity is the sudden transition to turbulence that the small surface boundary layer on a cylinder experiences at $Re \approx 2.5 \times 10^5$ (for smooth cylinders). Turbulent boundary layers have a different velocity profile shape than laminar boundary layers. The presence of turbulence enables momentum to be transferred across the shear layer convectively as well as viscously. This momentum transfer causes higher flow velocities to be present much closer to the cylinder surface in a turbulent boundary layer.

The increased momentum in a turbulent boundary layer enables it to remain attached to the cylinder surface much longer in the presence of the increasingly adverse pressure gradient that develops as the cylinder surface curves away from the flow direction on the downstream side. Prior to the boundary layer's transition to turbulence, the flow separates from a cylinder surface upstream of the centerline, whereas a turbulent surface boundary layer separates downstream of the cylinder centerline. This difference in separation points affects the width of the wake and the width of the negative pressure zone on the downstream side of the cylinder. This zone is narrower when the flow separation is delayed until after the boundary layer has passed the cylinder centerline, and consequently the aerodynamic drag is lower. Vertical vessels are large enough that the Re for realistic atmospheric flow conditions will be approximately equal to 10^7, which exceeds this critical regime. Smaller elements, such as small diameter pipes and cylindrical rail elements must be considered more closely.

Surface Roughness

Aerodynamic drag can be generated by two general mechanisms: form drag and friction drag. Form drag is caused by the development of pressures that act normal to the surface of a body, and this type of drag is typically dominant for bluff bodies such as cylinders. Friction drag is caused by the shear stresses parallel to the surface of a body that develop as a fluid passes over the body's surface. Friction drag has more importance for "streamlined" shapes, such as airplane wings, which do not produce large regions of separated flow (or wakes).

The presence of surface roughness on a cylinder can cause the surface boundary layer to become turbulent at lower values of Re than would be expected for a smooth cylinder surface. Consequently, the form drag would be lower for a rougher cylinder than for a smooth cylinder. However, there is a competing mechanism that mitigates this effect. The increased surface roughness increases the friction drag. So, even though lower drag coefficients can exist at lower values of Re, the ultimate drag coefficient at high Re is higher for rougher surfaces.

WIND LOADS FOR PETROCHEMICAL AND OTHER INDUSTRIAL FACILITIES 13

Aspect Ratio

The aspect ratio for a vertical vessel is defined as the height divided by the diameter. Reducing the aspect ratio for a circular cylinder causes a reduction in the aerodynamic force coefficient. The mean flow field for an infinitely long cylinder is two-dimensional. Finite length introduces three-dimensionality into the flow field. In the case of a finite-length cylinder, the flow has an alternative path around the cylinder – around the ends rather than solely around the circumference. This additional flow path reduces the quantity of flow that travels around the circumference on a per-length basis. The reduced quantity of flow translates into lower pressure on the circumferential surface of the cylinder.

For vertical vessels at grade, only the top end is effective in providing an alternative flow path. This orientation is similar to a chimney or a stack. *ASCE 7* provides force coefficients for these types of structures that consider the effects of finite length, and Chapter 5 of this document recommends the use of these values for vertical vessels. If a vessel is not oriented with one end at grade, then the use of force coefficients for ground-mounted cylindrical structures in *ASCE 7* will likely result in conservative estimates of the wind load.

Treatment of Appurtenances

Vertical vessels in real industrial settings are not ideal, smooth cylinders. Most vertical vessels have appurtenances such as ladders, handrails, platforms and piping of various diameters on their surfaces. The presence of these elements complicates the flow field and the calculation of wind loading. When the details of these elements are unknown (e.g. during preliminary analysis or design) it is suggested in the provisions of Chapter 5 of this document to account for their contribution to the wind load by augmenting the dimensions of the vessel itself. When more details are known, the contribution of these elements to the total wind load can be calculated by considering the projected areas and corresponding force coefficients of the elements. Considerable shielding will occur, and it is recommended in the provisions of Chapter 5 to only consider the net projected area of elements on the windward side of the vessel. An exception is for handrails on platforms, which can have sufficient separation to preclude shielding

Neighboring Vessels

It is common that vertical vessels are arranged closely as pairs or in other groupings. An example of such a pairing is a fluid catalytic cracking reactor and a catalyst regenerator. When vessels are closely arranged, the aerodynamics can be considerably more complicated. For some wind directions, an upstream vessel will shield the neighboring vessel, thus reducing the wind load on the downstream vessel. When vessels are oriented side-by-side with respect to approaching wind, their proximity can cause a local acceleration of the wind, thus increasing the wind loads on both structures. Chapter 5 suggests increasing the force coefficient for these crosswind arrangements. However, there is no provision in Chapter 5 to account for the possibility of shielding, which is consistent with the provisions of *ASCE 7*.

Directionality

The radial symmetry of single vertical vessels increases the likelihood that the worst wind direction will coincide with the critical aerodynamic orientation. However, for pairs or groups of vertical vessels, this radial symmetry is disrupted. In these scenarios, the designer may choose to use a lower value of the directionality factor, K_d.

Gust Effects

Vertical vessels can be quite slender, and they can be quite massive when filled with fluid. As such they may have lower fundamental vibration frequencies. The reduced natural frequency increases the likelihood that along-wind buffeting from the wind will amplify the wind loads. This effect is represented in the calculation of the gust effect factor, G. The example calculation in Chapter 6 illustrating the estimation of wind loads for vertical vessels demonstrates the approximate magnitude that this type of dynamic load amplification can produce. In this example, the calculated values of G for the rigid and flexible cases are 0.85 and 1.124, respectively, corresponding to a dynamic load amplification of approximately 32% when the vessel is filled with fluid.

2.6 Other Wind Loading Codes, Standards, and Guides

In the United States, *ASCE 7* governs the estimation of wind loads for analysis and design purposes (except for bridges, which are covered either by the AASHTO Standard Specification or AASHTO LRFD). *ASCE 7* is incorporated by reference into the International Building Code. This guide publication, *Wind Loads for Petrochemical and Other Industrial Structures*, is intended to supplement, and be compatible with, the wind load provisions of *ASCE 7*. However, the committee responsible for this document recognizes that it does not cover all the possible structure types or forms that may be encountered in the process industries. Members of this committee have found the following documents and resources useful, as they contain a wide variety of wind loading shape coefficients:

- AS/NZS 1170.2:2002, Australian/New Zealand Standard, Structural Design Actions, Part 2: Wind Actions.

- Eurocode 1: Actions on Structure – General Actions – Part 1-4: Wind Actions.

- ESDU Wind Engineering Series. ESDU (formerly the Engineering Sciences Data Unit) is a subscription based service that provides access to engineering data and software.

It is important to note that wind load provisions contained in the resources above may have a different basis than that of *ASCE 7* and this document. The analyst or designer should be careful to make sure that the use of various formulas and coefficients is compatible with other aspects of the analysis methodology (i.e. wind speed averaging times, mean versus peak effects, etc.).

2.7 Research Progress and Future Needs

The previous edition of this publication reviewed the analysis practices of several companies and provided consensus guide provisions for the estimation of wind loads. For the most part, the resulting provisions were based on extensions of methods from *ASCE 7* and engineering judgment rather than on research. Recognizing this limitation with the guide provisions, the committee identified several research priorities. Since then, progress has been made in addressing some of these issues.

- Force Coefficients and shielding effects for pipes in a pipe rack structure were studied in a wind tunnel by Liu, et al. (2008). The effects of shielding were found to be dependent on the number of cylinders in each pipe group, their spacing, and the turbulence in the flow. The wind tunnel experiments were in relatively low Reynolds numbers, so further research is required to more fully define the aerodynamics of pipe racks at full scale.

- Insight into the shielding of equipment inside open frame structures has been sought out in the literature and studied in a wind tunnel (Amoroso and Levitan, 2009a). It has been found that the force reductions for equipment inside these structures are generally greater than the provisions of the previous edition of this guide suggested.

- The question of an upper bound force coefficient for open frame structures has also been investigated, and it has been found that no such limit exists for unlimited plan geometries. However, if the envelope dimensions of an open frame structure are known, a reasonable upper bound force coefficient can be estimated (Amoroso and Levitan, 2009b).

- Limited experimental data are now available concerning the influence of a large diameter vertical pipe adjacent to a vertical vessel. It has been observed that aerodynamic interference between these two elements can sometimes amplify the wind loads (Amoroso, 2007).

- The effectiveness of simplified methods for estimating the contributions of ladders, nozzles, and small pipes to vertical vessel wind loads by using an increased vessel diameter has been evaluated for a particular case (Amoroso 2007 and Amoroso and Levitan, 2009a). It was found that the simplified method recommended by the previous edition of this publication was sometimes not conservative.

- The presence of solid flooring has been found to reduce the mean wind loads for open framed structures (Amoroso, 2009a).

- Petersen (1997) studied the wind environment for models of refineries in wind tunnel experiments. The motivation for the research was to improve the analysis of the mixing and dispersion of pollutants in the atmosphere. The experiments showed that the surface roughness inside the plants corresponded to suburban and dense urban terrain.

Despite this progress, many of the research questions identified in the first edition have not been addressed to the knowledge of this committee. These are:

- What are appropriate force coefficients and shielding effects for cable trays in a pipe rack structure?

- What are the loads on vertical rows of pipes (force coefficients for wind directions from perpendicular to parallel to the row of pipes)?

- What force coefficients should be used for different size, type, and orientation of platforms on a vessel?

- Although a limited case study has been performed, a general understanding of the appropriate force coefficients for ladders, nozzles, and small pipes remains outstanding. Can these items be accounted for as proposed in the Chapter 5 recommendations?

- What are the effects of irregular (nonrectangular) plan view open frame structures on the magnitude and application of the wind induced force?

In addition to the above outstanding research questions, discussions among committee members have identified the following items for consideration by researchers:

- How does the currently available research regarding pipe shielding compare with results from high Reynolds number experiments?

- What are the longitudinal wind forces on pipes in a pipe rack structure?

- Is the gust effect factor for rigid buildings as specified by *ASCE 7* applicable for rigid open frame structures?

- Although more refinement has been achieved in estimating the shielding effects for equipment inside open frame structures, the shielding estimates are gross quantities applied uniformly to all equipment items on a given story. What are the wind loads for individual pieces of equipment?

- What are the components and cladding pressures for partially clad structures?

- How do the recommendations for estimating wind loads for air coolers compare to wind tunnel test results?

CHAPTER 3
REVIEW OF EXISTING DESIGN PRACTICES

3.1 Introduction

The ASCE Petrochemical Committee responsible for the first publication of this document in 1997 conducted a review of design practices within the engineering community at that time (mid 1990s) to address the variable nature of such practices for industrial structures. Thirteen design practices for estimating wind loads on pipe racks, open structures, and pressure vessels were reviewed. The results of that effort were subsequently presented in the first edition.

Section 3.2 presents the results of an updated survey to complement the efforts of the previous committee.

Section 3.3 provides a discussion of design practices for additional structural types that were selected for inclusion in this publication, specifically pipe bridges, partially clad structures, tanks, steel stacks, cooling towers, and air coolers.

3.2 Survey of Existing Practices and Impacts of First Edition of Guidelines

The 1997 ASCE Task Committee report "Wind Loads and Anchor Bolt Design for Petrochemical Facilities" attempted to address the variable nature of design practices for industrial structures within the engineering community. The present committee conducted a web based survey of design engineers and engineering managers from large engineering consulting firms to assess the level of acceptance and effectiveness of the 1997 publication and to determine future document improvement needs.

The level of acceptance and effectiveness of the previous 1997 document can be represented by the following survey results:

i. 80% of the respondents were knowledgeable of the document

ii. 57% of the respondents referenced the publication in their current company design specifications.

In referencing the previous 1997 document, the company specifications have:

- modified the tributary areas of pipes on racks

- reduced wind loading on piping in a pipe rack due to multiple pipe rack levels

- used an alternative force coefficient for cable trays.

The committee wishes to thank those who responded to the survey. The information gathered has and will continue to guide this committee in its efforts to provide for the needs of the petrochemical and industrial engineering community. A table of the responses to each question and the desired enhancements for the document is covered in Table 3.1. The committee has attempted to address the desired enhancements conveyed through the survey results, and some of them are reflected in the chapters of this document.

Table 3.1 Survey Responses

Questions	Responses	
How familiar are you with the 1997 ASCE Task Committee report "Wind Loads and Anchor Bolt Design for Petrochemical Facilities" for application of wind loads?	Not familiar	21%
	Somewhat familiar	25%
	Very familiar	54%
Does your company specifications reference the 1997 ASCE Task Committee report "Wind Loads and Anchor Bolt Design for Petrochemical Facilities" for application of wind loads?	Unknown	14%
	No	29%
	Yes	57%
Does your firm reduce the wind loading on the piping in a pipe rack due to multiple pipe rack levels?	Unknown	23%
	No	65%
	Yes	12%
Does your firm consider longitudinal loading on pipe racks?	Unknown	19%
	No	15%
	Yes	67%
Does your firm use an upper limit for evaluation of open frame structures (e.g. consider the load on a fully clad structure with the same envelope shape, etc.)?	Unknown	26%
	No	56%
	Yes, Specify	19%
	Specified responses included the following comments: • Varies among projects - upper limit as fully clad. • Enclosed structure ($C_f = 1.3$). • Yes, not to exceed load of fully clad structure. • Fully clad structure with a $C_f = 1.3$.	
Would you rather have recommendations that require more detailed/complicated (and perhaps more accurate) analysis methods or easier-to-use estimates that may provide more conservatism?	More Detailed	8%
	Easier to Use	81%
	Other, please elaborate	12%
	Specified responses included the following comments: • Both, so the engineer can choose. • Keep it simple, but provide examples & guidance. • More accurate analysis methods.	

Are there specific recommendations or modifications that you feel should be made to the Wind Loads on Petrochemical Facilities publication or in the presentation of the recommendations? If so, please describe the modifications that you feel would make the recommendations better and more useable to the engineering community.	Unknown	19%
	No	15%
	Yes	67%
	i. 81% indicated a preference for easier to use methods that may provide more conservatism versus more detailed analysis or providing both options in the document. ii. 67% consider longitudinal loading on pipe racks. iii. More examples or structural types (see Note vi). iv. Guidance for shielding. v. Inclusion of design guidance for offshore structures. vi. List includes pipe bridges, structures with dense equipment, arrays of structures/adjacent structures, stacks, cooling towers, partially clad structures, one-bay structures, compressor sheds, fin fan type coolers, conveyers, elevated enclosures. vii. Recommendations be directed to all industrial and not specifically petrochemical facilities.	

WIND LOADS FOR PETROCHEMICAL AND OTHER INDUSTRIAL FACILITIES 21

How beneficial would new recommendations from the ASCE Petrochemical Committee be for the following structures?		Very	Somewhat	Not
	Arrays of structures/adjacent structures	62%	25%	12%
	Compressor sheds	41%	41%	18%
	Structures with dense equipment	64%	32%	4%
	One-bay structures	52%	26%	22%
	Stacks	58%	38%	4%
	Partially clad structures	55%	41%	5%
	Cooling towers	58%	33%	8%
	Fin fan type coolers	36%	50%	14%
	Pipe bridges	76%	20%	4%
Are there any other structure types that you feel should be evaluated by the ASCE Petrochemical Committee? Which specific structure types?	Specified responses included the following comments: • Miscellaneous pipe supports. • Equipment foundations (design and transfer of wind loads to foundations at grade). • Offshore structures. • More guidance on when shielding can be employed and limitations of shielding for wind loading. • Office/Admin Buildings and Control Room Structures. • Sloping conveyor structures, especially in the longitudinal direction. • Elevated enclosures where the floor and roof are exposed to wind.			

3.3 Existing Design Practices for Structures New To Second Edition

This section addresses the design practices for calculating the wind load on the following:

- Pipe Bridges
- Partially Clad Structures
- Tanks
- Steel Stacks
- Cooling Towers
- Air Coolers

3.3.1 Pipe Racks and Pipe Bridges

Most of the design practices have accepted the recommended guidelines (Section 5.1) to design the pipe rack structure as an open frame structure with additional loads for pipes and cable trays. It is the experience of this committee that pipe friction and anchor loads govern the longitudinal design loads of a typical pipe rack. However, the engineer should exercise judgment to ensure longitudinal wind loads are considered for unusual pipe rack configurations, for example, when there is a significant change in elevation of the piping at a road crossing.

There were no design guidelines to design pipe bridges in the previous edition. This edition recommends applying wind load on the bridge framing members in a manner similar to the application of wind load on open frame structures, and applying wind load on piping in a manner similar to that on piping in pipe racks.

3.3.2 Partially Clad Structures

Structures with partial cladding can take a variety of forms in the context of a petrochemical facility. Typical structure types include portal framed metal buildings with cladding only covering the roof and portions of the side and end walls and process towers with cladding only on some exterior sides. The former structure type is commonly a low-rise building and may be used as a shelter for compressors or other equipment. The latter type may include stair towers or portions of equipment structures that require some isolation from neighboring spaces, and provisions in Chapter 5 address the latter type.

Until recently, the wind load guidance available for these types of structures was limited to the open building provisions of *ASCE 7*. However, the provisions of *ASCE 7* are only applicable for estimating roof pressures. The 2010 Supplement to the 2006 Metal Building Systems Manual (Metal Building Manufacturer's Association) includes provisions for estimating lateral forces for open-sided, multiple bay, portal framed buildings. These provisions are based on recent research conducted at the Boundary Layer Wind Tunnel at the University of Western Ontario (Kopp et al., 2010).

3.3.3 Tanks

Overview

Most storage tanks located in petrochemical facilities are fabricated from steel and conform to one of the following documents:

API-650	Welded Steel Tanks for Oil Storage
API-620	Design and Construction of Large, Welded, Low-Pressure Storage Tanks
AWWA D-100	Welded Carbon Steel Tanks for Water Storage

WIND LOADS FOR PETROCHEMICAL AND OTHER INDUSTRIAL FACILITIES

UL-142	Steel Aboveground Tanks for Flammable and Combustible Liquids
NFPA 22	Standard for Water Tanks for Private Fire Protection
NFPA 30	Flammable and Combustible Liquids
NFPA 30A	Code for Motor Fuel Dispensing Facilities and Repair Garages

With respect to wind loads, the documents offer the following:

API-650

Section 5.2.1k. The design wind speed (V) shall be 120 mph (190 km/hr), the 3-second gusts wind speed determined from *ASCE 7*, Figure 6.1, or the 3-second gust design wind speed specified by the purchaser (this wind speed shall be for a 3-second gust based on a 2% annual probability of being exceeded). The standard goes on to state that the design wind pressures are to be in accordance with *ASCE 7* for wind exposure Category C. As an alternative, pressures may be determined in accordance with *ASCE 7* (exposure category and importance factor provided by purchaser) or national standard for the specific conditions for the tank being designed.

Uplift Forces (per API-650)

Section 5.2.1k. The design uplift pressure on the roof (wind plus internal pressure) need not exceed 1.6 times the design pressure P determined in Section F.4.1. It goes on to stipulate that windward and leeward horizontal wind loads on the roof are conservatively equal and opposite and therefore not included in these pressures.

Survey of literature indicates that roof sheet damage to tanks is common, however when this does occur, the failure is similar to an internal over pressure failure of a tank, where the roof to roof angle fails. This in turns allows the stiffing angle or wind girder to stay in place and thus the contents of the tank are usually not released. In the case of external floating roofs, experience has shown that due to unsymmetrical loadings on the roof, the floating roof may tend to shift. In a combined rain and wind event, there have been instances where this shift results in sufficient accumulation of rain water away from the drain to result in sinking of the roof (roof failure).

Wind Load on Tanks – (Overturning Stability)

API-650, Sections 5.11.1 thru 5.11.3 provides analytical methods to determine stability for tankage. This is primarily a concern for small tanks. Section 5.12 of API-650 provides prescriptive methods to calculate requirements for anchorage.

API-620

Section 5.4.1k. The design wind speed (V) shall be the 3-sec gust design wind speed (mph) determined from *ASCE 7* Figure 6-1 or the 3-sec gust design wind speed specified by the Purchaser. When wind is specified as measured by fastest mile the speed shall be multiplied by 1.2. For tank components exposed to wind up to 80 ft above ground, the design wind pressures normal to the tanks outside surface shall be the pressures below, multiplied by $(V/120)^2$. For tank components located more than 80 ft above ground, use *ASCE 7* to determine wind pressures.

Table 3.2 API Wind Pressures

Surface	Direction	Average Pressure lb. / ft.²	Maximum Pressure lb. / ft.²
Cylinder	Inward	16	31
Sphere	Inward or Outward	16	31
Dome or Cone Roof or Bottom	Outward	30	50

Alternatively, pressures may be determined in accordance with *ASCE 7* or a national standard for the specific conditions for the tank being designed.

Average wind pressure on the roof shall be used to design the roof to shell compression region and for overturning. Maximum wind pressure shall be used to design the roof and shell.

AWWA D-100-05

AWWA D-100-05 offers the following regarding wind loads:

Section 3.1.4 Wind load. Wind pressure shall be calculated by the formula

$$P_w = q_z\, G\, C_f \geq 30\, C_f \qquad \text{(AWWA Eq. 3-1)}$$

Where P_w = wind pressure applied to horizontal projected area, in pounds per square foot

G = gust-effect factor. The guest-effect factor shall be taken as 1.0 or may be calculated using the procedure given in *ASCE 7*. The calculated guest-effect factor shall be based upon a damping ratio of 0.05 and shall not be less than 0.85

C_f = force coefficient

WIND LOADS FOR PETROCHEMICAL AND OTHER INDUSTRIAL FACILITIES

q_z = velocity pressure evaluated at height z of the centroid of the projected are, in pounds per square foot.

q_z = $0.00256 \, K_z I V^2$ (AWWA Eq. 3-2)

Where K_z = velocity pressure exposure coefficient evaluated at height z of the centriod of the projected area (see Table 3)

z = height above finished grade

Section 3.1.4.1 Basic Wind speed. The basic wind speed shown in Figure 1 are based upon a 3-second guest speed at 33 ft. (10.1 meters) above ground and an annual probability of 0.02 of being equal or exceeded (50 year mean recurrence interval). In special wind regions, tanks may be exposed to wind speeds greater than those shown in Figure 1. In such cases, the basic wind speed shall be specified.

Section 3.1.4.2 Velocity pressure coefficients are provided for Exposure C and Exposure D in Table 3 of AWWA D-100-05. Exposure C shall be used unless otherwise specified. The velocity pressure coefficient shall be evaluated at height z of the centroid of the project wind area. For intermediate heights, use linear interpolation of the velocity pressure coefficients.

The remainder of the documents listed in the overview section does not address wind loads. Therefore the designer needs to exercise caution.

3.3.4 Steel Stacks

Most stacks located in petrochemical facilities are steel stacks. There are two basic types of steel stacks:

(1) Self-supporting stacks - together with the foundation would remain stable under all working conditions without additional support, and

(2) Guyed or braced stacks - not all external applied loads are carried by the stack shell and therefore guys or braces are provided to ensure stability.

ASME STS-1-2006 "*Steel Stacks*" and 1999 CICIND "*Model Code for Steel Chimneys*" provide the requirements and guidelines for design, fabrication, erection, and maintenance of steel stacks and their appurtenances. With respect to wind loads, steel stacks should be designed to resist the wind forces in both along-wind and cross wind directions. In addition, the variation of pressure along the circumference of the shell should be considered.

Along-Wind Loads: For wind loads along wind direction, ASME STS-1-2006 requires that wind loads on steel stacks be calculated as the sum of two components, one caused by a mean wind speed and the other caused by fluctuating wind gusts. The dynamic response is accounted for using a modified gust factor (similar to the gust factor for flexible structures as described in Section 6.5.8 of *ASCE 7*) depending on the natural frequency and the geometric properties of the steel stack. A similar along-wind load provision is included in the 1999 CICIND.

Per Section 4.3.3.1 of ASME STS-1-2006, the design wind loads along-wind direction is given by:

$$w(z) = \bar{w}(z) + w_D(z)$$

and

$$\bar{w}(z) = \frac{C_f q_z D}{12(1 + 6.8 I_{\bar{z}})}$$

$$w_D(z) = \frac{3z M_0}{h^3}\left[G_f(1 + 6.8 I_{\bar{z}}) - 1\right]$$

Where

$w(z)$ = total along-wind load on stack per unit height, lb.f/ft.

$\bar{w}(z)$ = mean along-wind load on stack per unit height, lb.f/ft.

$w_D(z)$ = fluctuating along-wind load on stack per unit height, lb.f/ft.

C_f = force coefficient given in Table I-5 of Mandatory Appendix I of ASME STS-1-2006 (or Figure 6-21 of *ASCE 7*)

q_z = velocity pressure, calculated using Equation (4-4) of ASME STS-1-2006 or Equation (6-15) of *ASCE 7-05*, $q_z = 0.00256 V^2 I K_{zt} K_z$

D = diameter of stack at elevation under consideration, in.

$I_{\bar{z}}$ = intensity of turbulence at height z, calculated per Mandatory Appendix I of ASME STS-1-2006 (or Section 6.5.8 of *ASCE 7*)

z = elevation under consideration, ft.

h = height of stack, ft.

M_0 = moment at the base of the stack due to $\bar{w}(z)$ loading, lb.f-ft.

G_f = gust effect factor, calculated per Mandatory Appendix I of ASME STS-1-2006 (or Section 6.5.8.2 of *ASCE* 7-05)

Crosswind Loads due to Vortex Shedding Loads: When stacks are subjected to a steady wind, the periodic shedding of vortices will cause swaying oscillations in a direction normal to that of the wind. If the vortex shedding frequency is resonant with the natural frequency of the stack, it may result in large vibrations. Section 5.2.2 (a) of ASME STS-1-2006 can be used to evaluate if vortex shedding needs to be considered. Example calculations in Nonmandatory Appendix E of ASME STS-1-2006 may be followed if needed. Per Section 5.2.2 (a) of ASME STS-1-2006, fatigue analysis must be considered due to vortex shedding loads under certain conditions.

Ovalling: In addition to transverse swaying oscillations, a steel stack may also be subjected to ovalling. Ovalling results from positive pressure on the up wind side of the stack and negative pressure on the sides and back. Ovalling resonance in the circular cross-sectional plane occurs as a result of vortex excitation. A lined stack is more resistant to ovalling because the lining contributes to high natural frequency and increased damping for the elastic ring. Therefore, ovalling need not be considered for lined stacks. Unlined stacks possess very little damping to restrict ovalling, and may experience excessive stresses and deflections at the critical ovalling wind velocity. Per Section 5.2.2 (b) of ASME STS-1-2006, it may be assumed that the unlined stack is prone to ovalling vibrations if the critical wind velocity for ovalling, as calculated in Equation (5-6), is less than or equal to mean hourly wind speed. A simple practical cure against ovalling is to use stiffening rings meeting the requirements of Table 4.4.7 of ASME STS-1-2006.

3.3.5 Cooling Towers

Most cooling towers located in petrochemical facilities are wood framed structures constructed on-site with corrugated cladding, wooden or fiber reinforced polyester (FRP) louvers, and FRP fan cylinders (or shrouds). They typically are at or near grade, less than 60 ft tall, rectangular in shape, and have a modest inward slope on two opposite sides. In some cases, cooling tower structural systems are constructed of materials other than wood to increase their durability (FRP or reinforced concrete) or to provide portability (metal). The Cooling Technology Institute (CTI) standard, *CTI Code Tower Standard Specifications for the Design of Cooling Towers with Douglas Fir Lumber* [CTI Bulletin STD-114 (96)],[1] provides design guidance for wooden cooling towers. With respect to wind loads, Section 12.0 *Design Data*, Paragraph 12.1 *Wind*, states the following:

> "Unless otherwise specified, wind pressure design shall be in accordance with *ASCE* 7-88, 1990. In design of the component portions of the structure, consideration of positive and negative pressures on windward and leeward surfaces shall be taken into account. Design shall provide for the maximum

---

[1] At the time of publication of this report, CTI Bulletin STD-114(07) had been released.

forces which would result from any wind direction. There shall be no reduction of wind force taken for the possible shielding effect of structures adjacent to the cooling tower. Design shall take into account the various geometric shapes with corresponding shape factors to be applied for wind force calculations.

The dry weight of the tower shall be used in determining uplift forces. Adequate anchorage to foundations or the supporting structure shall be provided in accordance with the specified or local building code."

Prior to the creation of this standard, unless specified otherwise by the owner, the manufacturer established the wind load criteria for a given tower design. In such instances it is not always clear how the design wind pressures were determined. For example, a manufacturer may have simply stated, "The basic design criteria shall be a 30 psf wind load. Closer examination of the application of such design values reveals that, although they may be suitable for calculating loads on the main structural frame (i.e., the MWFRS), these values often underestimate high local wind pressures acting along the corners and edges of the tower frame. Nonetheless, given the current CTI standard, it is now recognized that the *ASCE 7* standard is to serve as the design basis for determining wind loads on cooling towers. If applied properly, the CTI standard together with *ASCE 7* should enable the designer to account for global and local wind pressures acting on a cooling tower structure.

A recent paper by Daniel S. Kelly of Evapco titled *Wind Load Rated Packaged Cooling Towers* (CTI paper No. TP06-22, 2006 CTI Annual Conference, February 5 - 8, 2006) addresses special requirements for cooling towers per the International Building Code (IBC) 2003, *ASCE* 7-02, and the Florida Building Code (FBC). Kelly correctly states that the approach taken to calculate wind loads in the IBC and the FBC is based on the *ASCE 7* Standard. Kelly also provides examples for each case demonstrating how to calculate wind loads on a cooling tower, citing the methodology and terminology used in *ASCE 7*. The IBC 2003 adopted a methodology similar to *Method 1 – Simplified Procedure* of *ASCE* 7-02. Kelly lists some, but not all, of the requirements of Section 6.4 of *ASCE 7*, which establishes the basis for using Method 1. As noted by Kelly, *Method 2 – Analytical Procedure* of *ASCE 7* applies when:

1. IBC 2003 does not apply,
2. The building has a mean roof height of 60 ft or more (i.e., the top of the cooling tower is at a height of 60 ft or more), and
3. The building (i.e., cooling tower) does not have a site location for which channeling or buffeting of upwind obstructions warrants special consideration.

Section 6.4 of *ASCE 7* also lists additional caveats for applying Method 1, among them that the building (i.e., cooling tower) is:

- a simple diaphragm building,

- enclosed,
- regular shaped,
- not classified as a flexible structure,
- does not have response characteristics making it subject to across wind loading, vortex shedding, or instability due to galloping or flutter, and
- approximately symmetrical in cross-section in each direction with either a flat roof, or a gable or hip roof with 45 degrees.

The results as summarized by Kelly indicate that *ASCE 7* provides suitable guidance for the design of cooling towers. Given the structural features of most cooling towers, and to reduce the time required to calculate wind pressures, it is recommended that a cooling tower for a petrochemical facility be designed for wind loads using *Method 1 – Simplified Procedure* of *ASCE 7*. However, if a cooling tower is located on top of a building or other supporting structure in a manner that results in the top of the tower being at a height greater than 60 ft., then it is recommended that it be designed for wind loads using *Method 2 – Analytical Procedure* of *ASCE 7*.

Given the presence of the fan cylinder (or shroud) on top of the cooling tower, it is recommended that the calculation of wind pressures across its surface be treated separately from the calculation of wind pressures across the main cooling tower structure. The cylinder is similar in geometry and aspect ratio to open top tanks. The American Petroleum Institute (API) provides design specifications for such structures. Specifically, Section 3.2.1(f) of API-650 Addendum 4 states that for the vertical cylindrical portion of a tank the force will be as follows:

$$F_{API} = (18 \text{ psf}) (V/120)^2 (H) (D)$$

H and D are the tank height and diameter, respectively, in feet ($H \times D$ = vertical projected area), while V is the 3-sec gust wind speed at the site in mph.

Section 6.5.15 of *ASCE 7* provides the following equation [Eq. 6-28]:

$$F_{ASCE} = q \, G \, C_f A_f = (0.00256 \, K_z K_{zt} K_d V^2 I) \, G \, C_f A_f$$

Assuming $K_z = 0.98$ (Exposure C at 30 ft.), $K_{zt} = 1.0$, $K_d = 0.85$, $V = 120$, $I = 1.0$, $G = 0.85$, $C_f = 0.7$ (*ASCE 7* Figure 6-21), and $(H)(D) = A_f$, then $F_{API} = F_{ASCE}$. Therefore, to calculate the load on the fan cylinder (or shroud) the following equation is recommended based on *ASCE 7* Eq. 6-28:

$$p_z = (0.00256 \, K_z K_{zt} K_d V^2 I) \, G \, C_f = (0.00256) \, (K_z) (K_{zt}) (0.85) \, V^2 \, I \, (0.85) (0.7)$$

or

$$p_z = K_z K_{zt} (V/28)^2 I \text{ [psf; mph]}.$$

3.3.6 Air Cooled Heat Exchangers (Air Coolers or Fin-Fans)

Refinery and petrochemical industries have often installed air coolers as cooling facilities. At one time a specific manufacturer called their air coolers "fin-fans" and the term later came to be a reference to any air cooler. For the convenience of piping access and conservation of plot space, the most common location to install air coolers is at the top of new or existing pipe racks. Most air coolers have maintenance access platforms.

For the estimation of wind loads, air coolers have been treated similarly to buildings at grade. However, since air coolers are elevated, in the opinion of this committee they may be considered as isolated blocks which allow air to flow beneath them as well as over and around them. A method to estimate the wind load on elevated blocks is presented in Section 5.6.

CHAPTER 4
RECOMMENDED GUIDELINES PART I: DESIGN CONSIDERATIONS AND METHODS

4.0 Introduction

Chapter 4 provides guidance for wind load analysis and design of petrochemical and industrial structures, and includes a broad range of topics related to wind effects for these types of facilities.

This chapter begins with observations on the historical performance of petrochemical facilities during hurricanes and provides some related general design considerations. Sections 4.2 and 4.3 provide background on alternate methods for the determination of wind loads, specifically wind tunnel testing and computational fluid dynamics approaches. Information is provided in Section 4.4 regarding recommended wind load combinations for petrochemical structures. Special wind load provisions related to United States Department of Transportation regulations for Liquefied Natural Gas (LNG) facilities are presented in Section 4.5. Section 4.6 provides references and summaries of current industry guidelines for the evaluation of wind loads on existing structures and foundations. Finally, a discussion of the uncertainty associated with the determination of wind loads is provided in Section 4.7.

4.1 Historical Performance

This section highlights the performance of industrial structures during recent hurricane events. It should be noted that comprehensive performance data are not publicly available. It should also be noted that during the last 40 years, many industrial facilities along the Gulf and Atlantic coastlines have experienced at least a tropical storm or a minor hurricane and seemed to have performed well. However, many of these facilities have not experienced a design level event (typically a moderate Category 2 or stronger hurricane), thus leading some in industry to believe that their facilities can withstand major hurricane force winds without sustaining significant damage.

A recent paper by Godoy, titled "Performance of Storage Tanks in Oil Facilities Damaged by Hurricanes Katrina and Rita" (ASCE Journal of Performance of Constructed Facilities, November/December 2007), indicates that performance of oil storage tanks was inconsistent. The paper points out that floating roof tanks with wind girders in place fared well, while adjacent similar tanks without wind girders suffered significant damage.

The poor performance of electrical power transmission and distribution systems can be a major problem at an industrial facility. A recent paper by Calvert and Fouad, "A Review of Current Wind Load Provisions for Transmission Pole Structures, ASCE Structures 2000," points out that the National Electrical Safety Code (NESC) wind loads applied to power poles are significantly lower than those calculated by *ASCE 7*.

Provided below are observations of damage sustained during some past hurricane events. In each case a storm category is provided at landfall per the National Hurricane Center, however, the wind intensity at the site is unknown.

Hurricane Celia – August 1970 (Category 3)

The photograph below is of a damaged catalytic unit at a refinery in Corpus Christi following Hurricane Celia in 1970. According to a (blurry) sketch of the unit prior to the event, the unit collapsed at a distance about 1/3 of its height from its base, with a surge separator, hot bins, and a steam drum toppling to the ground. The unit was noted as being one of the older structures on the site.

Figure 4.1.1 Corpus Christi Refinery Unit – Hurricane Celia

Hurricane Hugo - September 1989 (Category 4 at St. Croix)

Hurricane Hugo passed through the Caribbean Region leaving billions of dollars of damage in its wake. Amerada Hess incurred significant expenses repairing and upgrading its St. Croix Refinery, Virgin Islands, after it sustained heavy damages from the storm. The National Research Council and J&H Marsh and McLennan reported that a large number of petroleum tanks, including 14 of the 500,000 to 600,000 barrel capacity storage tanks sustained extensive damage at the refinery.

WIND LOADS FOR PETROCHEMICAL AND OTHER INDUSTRIAL FACILITIES 33

Figure 4.1.2 Petroleum Storage Tanks – Hurricane Hugo
(Photo Courtesy of NOAA)

Hurricane Charley – August 2004 (Category 4)

Hurricane Charley was a compact storm that produced design level winds near its landfall location along the southwestern coast of Florida. It was the most intense hurricane to make landfall in the United States since Hurricane Andrew in 1992.

Figure 4.1.3 Building with Masonry Infill – Hurricane Charley

34   WIND LOADS FOR PETROCHEMICAL AND OTHER INDUSTRIAL FACILITIES

Hurricane Frances - September 2004 (Category 2)

Hurricane Frances made landfall along the Florida east coast. It was one of four storms that impacted Florida in 2004. Some buildings were in a temporary state of repair when Frances hit. Prolonged power outages were also experienced during this time.

Figure 4.1.4 Power Pole – Hurricane Frances
(Photo Courtesy of NASA KSC)

Figure 4.1.5 Cable Tray Damage – Hurricane Frances
(Photo Courtesy of NASA KSC)

*Figure 4.1.6 Cladding Damage – Hurricane Frances
(Photo Courtesy of NASA KSC)*

Hurricanes Katrina and Rita – August-September 2005 (Category 3)

During Hurricanes Katrina and Rita, power supplies were severely affected, as approximately 1,000,000 wooden poles and approximately 300 towers or steel poles collapsed due to wind forces. Several steel oil storage tanks also were severely damaged due to wind forces, while numerous other tanks were damaged by flooding after floating away from their original positions. Several buildings suffered failure of exterior load bearing and non-load bearing masonry walls. Many cooling towers sustained damage of components and cladding, and loss of fan shrouds. Insulation and lagging systems suffered moderate to severe damage, and this damage did not seem to be limited to just one system type. Forensic data suggests that pre-existing conditions played a significant role in the damage on all types of structures.

36 WIND LOADS FOR PETROCHEMICAL AND OTHER INDUSTRIAL FACILITIES

*Figure 4.1.7 Storage Tank under Construction – Hurricane Katrina
(Photo Courtesy of Louisiana Chemical Association)*

*Figure 4.1.8 Oil Storage Tanks – Hurricane Katrina
(Photo Courtesy of NOAA)*

*Figure 4.1.9 Tank Damage - Hurricane Katrina
(Photo Courtesy of NOAA)*

*Figure 4.1.10 Cooling Tower Damage – Hurricane Katrina
(Photo Courtesy of Louisiana Chemical Association)*

38   WIND LOADS FOR PETROCHEMICAL AND OTHER INDUSTRIAL FACILITIES

*Figure 4.1.11 Cooling Tower Damage- Hurricane Katrina
(Photo Courtesy of Louisiana Chemical Association)*

*Figure 4.1.12 Tank Cladding Damage – Hurricane Rita
(Photo Courtesy of Louisiana Chemical Association)*

*Figure 4.1.13 Tank Insulation and Cladding Damage – Hurricane Rita
(Photo Courtesy of Louisiana Chemical Association)*

*Figure 4.1.14 Flare Tower Damage – Hurricane Rita
(Photo Courtesy of FEMA)*

40 WIND LOADS FOR PETROCHEMICAL AND OTHER INDUSTRIAL FACILITIES

Figure 4.1.15 Utility Pole - Hurricane Rita
(Photo Courtesy of FEMA)

Figure 4.1.16 Power Poles – Hurricane Rita
(Photo Courtesy of NIST)

*Figure 4.1.17 Electrical Transmission Tower – Hurricane Rita
(Photo Courtesy of NIST)*

*Figure 4.1.18 Cooling Tower Damage – Hurricane Rita
(Photo Courtesy of Louisiana Chemical Association)*

*Figure 4.1.19 Cooling Tower Damage – Hurricane Rita
(Photo Courtesy of Louisiana Chemical Association)*

*Figure 4.1.20 Crane Boom across Pipe Rack – Hurricane Rita
(Photo Courtesy of Louisiana Chemical Association)*

Hurricane Humberto – September 2007 (Category 1)

Hurricane Humberto was probably the fastest developing storm in recent time. The storm rapidly intensified from a tropical depression into a hurricane within 19 hours. The storm inflicted more damage than its Category 1 strength would suggest. It is reported that the storm caused damage to an LNG storage vessel in early construction phase at the Texas-Louisiana coast line. It is suspected that the tank was vulnerable due to the phase of construction underway at the time of the event. In addition to the industrial equipment damage, there were numerous power outages that affected several industrial complexes.

4.2 Wind Tunnel Testing

ASCE 7 allows the use of wind tunnel testing to estimate wind loads in lieu of the simplified and analytical procedures. The recommended provisions in this guide are classified with the analytical procedure for "other structures." The use of the analytical procedure is limited, and *ASCE 7* Section 6.5.2 requires that structures be designed using either recognized literature documenting the load effects or wind tunnel testing in cases where:

- a structure may have across-wind loading,
- vortex shedding may occur,
- instabilities due to galloping or flutter are possible,
- a structure has a site location for which channeling effects or buffeting in the wake of upwind structures warrant special consideration,
- a structure has unusual shape, or
- a structure has unusual response characteristics.

In addition to these limitations on the analytical procedures, there are other considerations particular to petrochemical structures that may make wind tunnel testing an attractive alternative method:

- A structure is not adequately covered by the provisions of this document. This limitation is similar to the limitation above regarding unusually shaped structures.
- It is suspected that the currently available wind load analysis methods may be overly conservative, and a more precise understanding of wind loads may result in a more efficient structure. For example, the analytical procedure explicitly prohibits reductions in velocity pressure due to apparent shielding from neighboring structures.
- The adequacy of an existing structure to support new or modified equipment or appurtenances is being investigated.
- A validation of wind load analysis practices is desired.

If a wind tunnel test is to be used to estimate wind loads or wind load response for a structure, it is important that the simulation properly represents the velocity profile and turbulence characteristics of the atmospheric boundary layer; the structures and the surrounding topography are geometrically similar to the prototype; wind tunnel blockage is limited; wind tunnel pressure gradients are accounted for; Reynolds number effects are minimized; and the response characteristics of the instrumentation are consistent with the planned measurements. These requirements are listed in *ASCE 7* Section 6.6.2, and more detailed guidance on proper wind tunnel testing practice is contained in ASCE Manual of Practice No. 67 (ASCE, 1999). It should be noted that a new ASCE Standard on Wind Tunnel Testing is nearing completion, which will supersede the Manual of Practice. Particular care should be taken in minimizing Reynolds number effects. Due to the presence of exposed framing elements, equipment, and appurtenances that are common with petrochemical structures, length scales far below what is typical for enclosed buildings will be present. This complicates the determination of characteristic length for dynamic similarity considerations.

There are three common model types used in the wind tunnel testing of civil engineering structures: rigid pressure models, force balance models, and aeroelastic models. Rigid pressure models are used to measure surface pressures on building models. Small holes on the surfaces of these models transmit surface pressures to pressure transducers via thin tubing. Pressure model tests are particularly useful for determining the peak pressure coefficients that control the design of components and cladding with small tributary areas. As the price of electronic instrumentation falls, simultaneous measurement and integration of surface pressures is becoming a more common technique for estimating overall building loads.

Force balance models are typically used for determining the overall force, moment, and torsion coefficients for a particular structure. A force balance is an instrument consisting of load cells and/or strain gauges. When the fundamental frequency of the model/force balance system is high, this technique can be used to measure the overall fluctuating forces and moments generated by the interaction of the turbulence in the approaching wind and the structure geometry. The fluctuating forces can be used in conjunction with the structure's response mode shapes and frequencies to estimate the dynamic response of the prototype in natural winds.

Aeroelastic models are used for structures that are so flexible that there is a significant coupling between the aerodynamic forces and the dynamic response of the structure. A proper aeroelastic model correctly represents the mass, stiffness, and damping characteristics of the prototype structure at model scale. These are the most expensive and complicated wind tunnel tests, and are commonly used for structures such as long-span bridges.

At the time of the publication of this document, schedules for typical wind tunnel testing programs ranged from weeks to months, and the cost of commissioning such a study ranged from several thousand to many tens of thousands of dollars.

4.3 CFD Commentary

Determination of wind loads on many types of structures, including some of the structures that make up a petrochemical facility, is based in large part on the results of wind tunnel testing. Such tests are conducted either on full-scale component members, or on scale models of actual structural configurations. Although wind tunnel testing is a recognized and well-established method of determining wind loads, in select cases computational fluid dynamics modeling (CFD) may be a suitable alternative method for evaluating wind loads on structures.

CFD is a technique that uses numerical models to compute fluid flow around or through a body. When applied properly, the benefits are obvious. By understanding the effects of wind flow acting on a structure early in the design process, designers can alter the structural configuration to optimize performance. For this reason the automotive industry has invested heavily in CFD models to accelerate development of new vehicle models by applying CFD models early in the design process.

Although modeling fluid flow is still an evolving science, CFD models have advanced considerably since their introduction in the 1970s. Most CFD models require solving a highly non-linear set of partial differential equations known as Navier-Stokes equations. Used to describe fluid motion mathematically, the Navier-Stokes equations provide a relationship between density, pressure, viscosity, and velocity that conserves mass and momentum. However, they cannot be solved exactly except for special cases. Hence, the standard approach to obtain a solution is to divide the flow domain into a large number of small cells, then calculate the fluid parameters on a rigid grid. For more complex problems, this approach can cause physical and numerical instabilities that, in turn, can cause calculations to abort. Since these equations sometimes experience numerical instabilities, sophisticated numerical techniques are required to achieve a stable solution that reasonably matches experimental results.

CFD models using Navier-Stokes equations assume that the fluid is a continuum material, rather than a collection of molecules. Another CFD approach is to use what physicists refer to as the "lattice Boltzmann method" in place of the Navier-Stokes equations. A CFD model based on the lattice Boltzmann method is basically a discrete particle system having a finite number of states that are updated in distinct steps. Each step follows either deterministic or stochastic rules dictated by the conservation of mass, momentum, and energy. The resulting behavior is equivalent to, but independent of, continuum mechanics. An intriguing aspect of the lattice Boltzmann method is that the solution is identical to the solution produced by Navier-Stokes equations, yet no differential equations are being solved by the process.

To be sure, all CFD models have had mixed results in reproducing the separation and the reattachment of fluids at high wind velocities when the Reynolds number is large and the flow is turbulent. Estimations of mean pressures in zones of separation and reattachment are hence more likely to be suspect, while errors in predicting mean and fluctuating surface pressures may also lead to errors in predicting overall wind loads for certain geometries. Another challenge from a computational standpoint is the complexity of a three-dimensional geometry associated with structures like an open frame process tower that is made up of a collection of round and sharp-edged members. However, rapid advancements in computer designs and operating system development have made this challenge less daunting.

When determining wind loads on structures that make up a petrochemical facility, cases may arise that justify the use of a CFD model. In such cases careful consideration should be given to ensuring that the CFD model is properly applied during analysis of the structure or component. When using a commercially available CFD program, request that the vendor provide case studies of the application of their model to wind-related analyses, including benchmarking to wind tunnel or full-scale measurements. Also, be mindful of the computing requirements to execute the CFD runs. Finally, recognize that more complex geometries involving a multitude of assemblies may prove too difficult to solve in practice due to computing hardware limitations. In general the cost of CFD modeling is comparable to wind tunnel testing. Although CFD modeling allows review of wind effects early in the design process, this benefit is offset to a degree by the limited usefulness of such results when compared to wind tunnel or full-scale testing.

4.4 Load Combinations

This section addresses load combinations that include the effects of wind loading. Other load combinations that are absent of wind loading are not considered in this guideline.

The engineer is responsible for verifying applicable codes and standards by state or local agencies that have jurisdiction in the state where the project is constructed.

4.4.1 References to Existing Industry Guidelines

Building codes, standards, and industry guidelines address load combinations for buildings and structures. This section highlights some of the main industry standards used for loading and load combinations; it is not meant to be an exhaustive list. The following references discuss load combinations in the presence of wind loading.

4.4.1.1 ASCE 7

The basic load combinations set forth in Section 2 of *ASCE 7* are the minimum to be considered for the design of buildings and structures. Load combinations 3, 4, and 6 of Section 2.3.2 and combinations 5-7 of Section 2.4.1 are applicable for addressing the effects of wind loads in combination with other loads.

WIND LOADS FOR PETROCHEMICAL AND OTHER INDUSTRIAL FACILITIES   47

Strength Design from Section 2.3.2

$1.2D + 1.6(L_r \text{ or S or R}) + (L \text{ or } 0.8W)$ (3)
$1.2D + 1.6W + L + 0.5(L_r \text{ or S or R})$ (4)
$0.9D + 1.6W + 1.6H$ (6)

Note: The load factor on L in load combination 3 and 4 is permitted to equal 0.5 for all occupancies where the minimum distributed live load in Table 4-1 of *ASCE 7* is less than or equal to 100psf, with the exception of garages or areas occupied as places of public assembly.

Allowable Stress Design from Section 2.4.1

$D + H + F + (W \text{ or } 0.7E)$ (5)
$D + H + F + 0.75(W \text{ or } 0.7E) + 0.75L + 0.75(L_r \text{ or S or R})$ (6)
$0.6D + W + H$ (7)

Where:
D = Dead Load
E = Earthquake Load
F = Load due to fluids
H = Load due to pressure (lateral earth, ground water, bulk material)
L = Live load
L_r = Roof live load
R = Rain Load
S = Snow Load
W = Wind Load

In accordance with load combination 6 above, it is permitted to multiply variable loads by 0.75 if two or more variable loads are considered to occur simultaneously in the same load combination.

Load combination 6 in Section 2.3.2 and load combination 7 in Section 2.4.1 address effects due to loads counteracting gravity. One of the important design considerations for these combinations is the stability of structures. In V-Zones and Coastal A-Zones, where simultaneous hurricane wind speeds and storm surge is possible, Section 2.3.3 needs to be examined. This section states that 1.6W and 2.0Fa shall be considered in these regions. The high load factor for flood loads results from variability and uncertainty in flood conditions in comparison to other environmental loads.

4.4.1.2 International Building Code (IBC 2009)

The load combinations in Section 1605 of the International Building Code correlate with those set forth in *ASCE 7*. The following wind load combinations are to be considered for design:

Strength Design

$1.2D + 1.6 (L_r \text{ or } S \text{ or } R) + (f_1 L \text{ or } 0.8W)$	(IBC Eq. 16-3)
$1.2D + 1.6W + f_1 L + 0.5 (L_r \text{ or } S \text{ or } R)$	(IBC Eq. 16-4)
$0.9D + 1.6W + 1.6H$	(IBC Eq. 16-6)

Allowable Stress Design

$D + H + F + (W \text{ or } 0.7E)$	(IBC Eq. 16-12)
$D + H + F + 0.75 (W \text{ or } 0.7E) + 0.75L + 0.75 (L_r \text{ or } S \text{ or } R)$	(IBC Eq. 16-13)
$0.6D + W + H$	(IBC Eq. 16-14)

Where:
D = Dead Load
E = Earthquake Load
F = Load due to fluids
f_1 = 1 for floors in places of public assembly, for live loads in excess of 100 psf and for parking garage live loads, and 0.5 for all other live loads
H = Load due to lateral earth pressure, ground water pressure, or pressure of bulk materials
L = Live load
L_r = Roof live load
R = Rain Load
S = Snow Load
W = Wind Load

In accordance with Equation 16-13, it is permitted to multiply variable loads by 0.75 if two or more variable loads are considered to occur simultaneously in the same load combination.

IBC also provides alternate load combinations for allowable stress design in Section 1605.3.2 as shown below:

$D + L + \omega W$	(IBC Eq. 16-17)
$D + L + \omega W + 0.5S$	(IBC Eq. 16-18)
$D + L + S + 0.5\omega W$	(IBC Eq. 16-19)

Where ω is a wind load factor whose value is dependent on whether the directionality factor is utilized in the wind load calculation. When the wind load is determined per *ASCE 7*-05, the directionality factor, K_d is included as a factor in the calculation and $\omega = 1.3$. When wind loads are calculated other than per *ASCE 7*, then $\omega = 1.0$. If used for checking the counteracting effects of wind and dead load, only 2/3 of the minimum dead load at the point being considered shall be utilized. For these alternate load combinations, allowable stresses are permitted to be increased by the material chapter of the IBC or by referenced standard.

4.4.1.3 Process Industry Practices (PIP) Structural Design Criteria, STC01015 February 2006 (recently revised 9/07)

Since *ASCE 7* does not address loads and load combinations particular to the petrochemical industry, Process Industry Practices (PIP) Structural Design Criteria STC01015 was developed to assist engineers in developing consistent practices for the design of petrochemical facilities.

In general, PIP Section 4.2 is based upon the combinations set forth in *ASCE 7*. However, loads and load combinations for particular petrochemical structures and equipment are addressed. It should be noted that the allowable stress wind load combinations for uplift deviates from *ASCE 7* for equipment and pipe racks. These combinations utilize a dead load factor of 1.0 versus the 0.6 in *ASCE 7*. PIP's philosophy for variance from *ASCE 7* is that, in the petrochemical industry, weights for equipment and pipe racks are known and therefore a dead load factor of 1.0 is justified. When utilizing the PIP combinations for uplift, which utilize a 1.0 load factor for dead load, the foundation design should be checked against a stability ratio of 1.5 for overturning.

The engineer should be aware that use of the PIP load combinations, especially those for empty condition, may not comply with the requirements of *ASCE 7*. Use of the PIP load combinations should be discussed with the local building code official.

4.4.2 Recommended Guidelines for Petrochemical Facilities

4.4.2.1 General

ASCE 7 and other building codes typically do not address loading or load conditions applicable to petrochemical facilities. It is the attempt of this guideline to create petrochemical loads and load combinations which are aligned with *ASCE 7*, for consistent use across the industry for new structures and in the evaluation of existing structures.

Owner's and local building authorities' requirements should be consulted and may be different than those discussed in this guideline.

4.4.2.2 Loads

Loads for petrochemical facilities are unique in the way that they vary over time. For instance, dead load can be fairly complex in that a specific piece of equipment may have four to five weight conditions, depending on the stage of construction or operating condition.

Similarly, wind loads can vary over different loading conditions. For new facilities, under normal operating conditions, maximum wind loads should be utilized. However, for temporary conditions, such as construction and hydrostatic test, a reduced wind load may be considered.

The following load definitions are to be considered for use:

Definition of Loads

D_s Dead load due to self weight of structure, including foundation weight, soil overburden and fireproofing, as appropriate for the load condition being considered. Fireproofing is to be considered applicable for the structure and load condition considered. Field applied fireproofing should not be considered for the erection load condition, whereas shop applied fireproofing would be applicable in all load conditions.

D_e Dead load due to the empty weight of piping, electrical trays and equipment. Many times, the weight of piping associated with a vessel is estimated as a percentage of the empty weight of a vessel. Caution must be exercised not to over-estimate these loads for the empty or uplift conditions.

D_f Dead load due to fabricated weight of equipment, prior to appurtenances such as platforms, ladders and internals that are added during the construction.

D_o Dead load due to operating weight of piping, electrical trays and equipment (including fluid)

D_t Dead load due to test weight of piping and equipment

D_u Dead load due to upset load conditions for vessels during operation. Upset loads typically exceed normal operating loads.

D_h Dead load of equipment per the facility's hurricane procedures to be considered for evaluation of existing structures

L Live loads from platforms, equipment surge loads, and impact loads

T_a Thermal loads due to pipe anchor or guide loads, exclusive of friction loads.

W Wind load per *ASCE 7*, appropriate for the project location

W_{NH} Applicable to hurricane prone regions only. Recommend use of 90 mph as the non-hurricane basic wind speed.

W_p Partial wind load for use in short term conditions such as hydro testing of equipment, construction, or turnaround.

Per ASCE 37-02, in hurricane prone regions two options exist for partial wind load:

1. The partial wind load should be the *ASCE 7* design wind speed, adjusted by the appropriate factor in section 6.2.1 of ASCE 37-02, dependent on the estimated duration of construction. These factors range from 0.75 for a 6 week duration to 0.9 if duration is anticipated to be 2 to 5 years. If the construction is known to occur outside of hurricane season, then a basic wind speed of 90 mph may be used.

2. If the construction is known to occur during hurricane season, then a basic wind speed of 90 mph may be used with the appropriate factor from section 6.2.1 of ASCE 37-02, provided additional bracing is prepared in advance and applied prior to hurricane landfall.

Caution: Even when the construction is estimated to occur outside of hurricane season, the engineer should consider bracing provisions as a potential mitigation for schedule slippage that may shift construction into the hurricane weather window.

For extremely short test periods (days), consideration can be made to place a limit on environmental conditions when the test may be performed. Doing so may mean delay of testing, increasing construction costs and schedule, and should be done with caution.

For short-term conditions, use of the Importance Factor, $I = 1.0$ can be utilized.

4.4.2.3 Load Combinations for New Structures

Load combinations for new structures need to take into consideration the full time history of the facility, from construction through long-term operation. Load conditions such as erection, empty, and hydrostatic test that are common for petrochemical facilities are not included in *ASCE 7* and are addressed in Table 4.4.2.3.

Table 4.4.2.3 Minimum Wind Load Combinations for New Structures

Load Combination Description	Allowable Stress Load Combination	Strength Design Load Combination	Comments
Normal operation	$D_s + D_o + T_a + W$	$1.2(D_s + D_o + T_a) + 1.6W$	
Normal operation with live load	$D_s + D_o + T_a + 0.75(L + W)$	$1.2(D_s + D_o + T_a) + 0.5L + 1.6W$	
Erection	$0.6(D_s + D_f) + W_p$	$0.9(D_s + D_f) + 1.6W_p$	If construction is known to occur during hurricane season and a basic wind speed of 90mph is used in determining Wp, then temporary wind bracing must be provided.
Empty	$0.6(D_s + D_e) + W$	$0.9(D_s + D_e) + 1.6W$	
Test	$D_s + D_t + W_p$	$1.2(D_s + D_t) + 1.6W_p$	
Upset	$D_s + 0.75D_u + T_a + 0.75W_{nh}$	$1.2(D_s + T_a) + 1.0D_u + 1.6W_{nh}$ or $1.2(D_s + T_a) + 1.6D_u + 0.8W_{nh}$	Typically upset loads are due to process surge forces which may be considered as a live load.

Notes:
1. Engineering judgment is required to determine appropriate and critical load combinations.
2. Use of a 1/3 stress increase for allowable stress load combinations involving wind loads should not be used.
3. When performing foundation design, it is suggested that in the design for the Erection and Empty condition, with use of the 0.6 factor on dead loads, a stability ratio greater than or equal to 1.0 should be provided. The effects of buoyancy should be considered. For all other loading conditions with wind loading, a minimum stability ratio of 1.5 should be used.

4.4.2.4 Load Combination Considerations for Existing Structures

General

Modification to existing structures is often a balance between the reduction of risk and the cost of modification. In an effort to maximize the benefit and minimize the cost of an upgrade, most facility owners are receptive to practical design considerations. The engineer should contact the owner in the early stages of design

WIND LOADS FOR PETROCHEMICAL AND OTHER INDUSTRIAL FACILITIES

to make recommendations and agree upon special load combination provisions for the evaluation of existing structures. Once alignment with the owner is achieved in the areas discussed below, consultation and agreement from the local building officials may be required where the facility falls under local jurisdiction. The term "existing structures" is intended to include process structures, shelters, pipe racks and equipment. The purpose of these recommendations is to provide the engineer and owner with options in developing the appropriate level and cost of modifications.

Hurricane Event Considerations

In new structure design, hurricane wind speeds are utilized for normal operating conditions, regardless of the facility's hurricane response plan. Unlike seismic events, hurricane events are known in advance and the facility typically has a hurricane response plan that includes shutdown procedures. Facilities are not normally left operating during a hurricane and the engineer should consult with the owner for their site specific Hurricane Response Procedures. If well documented hurricane shutdown procedures exist, then the following items should be considered for the evaluation and modification of existing structures with the owner's agreement.

Operational Gravity Loads

Based on the facility hurricane procedures, vessels and tanks may be left with fluids inside to resist overturning forces from a hurricane event, while some may have inventories reduced. Depending on inventories and chemistry, tanks may be water-filled. The owner should be consulted on anticipated weights for major equipment and items during hurricane conditions, based on the hurricane shutdown plan. For evaluation of existing structures under hurricane wind speeds, the appropriate dead load of equipment per the facility's hurricane procedures should be utilized.

Live Loads

During hurricane events where the facility is in shutdown mode, non-essential personnel are evacuated and only essential staff remains. In this case, live loads would be minimal or non-existent and may therefore be reflected in the load combinations. Typically, the highest live loads are during construction, turn-around and maintenance activities due to personnel and temporary loads. In some instances, operational conditions, such as catalyst staging areas create high localized live loads. However, these activities would be performed in minimal wind conditions, significantly less than used for design purposes. Therefore, reduced live load should be considered for normal wind load combinations for modification to existing structures.

Friction Loads

Unless otherwise defined by hurricane shutdown procedures, since the plant will be in shutdown mode, friction loads should not be considered during hurricane conditions. As appropriate, cold spring and/or anchor loads may need to be considered. Any special hurricane load/load combination design basis and assumptions utilized for the evaluation of existing structures should be clearly documented on structural and process drawings and clearly indicated in the hurricane response plan and the facility shutdown procedures.

The recommended load combinations for evaluation of existing structures are provided in Table 4.4.2.4.

Table 4.4.2.4 Wind Load Combinations for Evaluation of Existing Structures

Load Combination Description	Allowable Stress Load Combination	Strength Design Load Combination	Comments
Normal operation	$D_s + D_o + T_a + W_{NH}$	$1.2 (D_s + D_o + T_a) + 1.6 W_{NH}$	For non-hurricane prone regions, use W in lieu of W_{NH}
Normal operation with live load	$D_s + D_o + T_a + 0.75 (L + W_{NH})$	$1.2 (D_s + D_o + T_a) + 0.5L + 1.6 W_{NH}$	For non-hurricane prone regions, use W in lieu of W_{NH}
Hurricane or Empty Condition	$D_s + D_h + W$ $0.6 (D_s + D_h \text{ or } D_e) + W$	$1.2 (D_s + D_h) + 1.6W$ $0.9 (D_s + D_h \text{ or } D_e) + 1.6W$	Structure design (other than covered below for foundations)
Hurricane or Empty Condition	$0.9 (D_s + D_h \text{ or } D_e) + W$		For checking foundation stability and pile tension (see Note 4)
Test	$D_s + D_t + W_p$	$1.2 (D_s + D_t) + 1.6 W_p$	
Upset	$D_s + D_u + T_a + W_p$	$1.2 (D_s + D_u + T_a) + 1.6 W_p$	
Notes:			
1.	Engineering judgment is required to determine appropriate and critical load combinations.		
2.	Use of a 1/3 stress increase for allowable stress load combinations involving wind loads should not be used.		
3.	The above load combinations should be discussed and approved by owner and local building official (if applicable) prior to implementation.		
4.	When performing foundation design, a stability ratio greater than or equal to 1.5 should be provided. Effects of buoyancy should be considered.		
5.	Operation Procedures must clearly denote that deviation from the hurricane response plan may compromise structure/facility integrity and must be reviewed by Engineering.		
6.	For the empty case, if it is determined that the structure or equipment will not be empty during the hurricane season, use of W_{NH} may be considered with Owner's approval.		

WIND LOADS FOR PETROCHEMICAL AND OTHER INDUSTRIAL FACILITIES

4.5 Special Considerations for LNG Facilities

4.5.1 Introduction

In recent years, there has been increased activity in the construction of Liquefied Natural Gas (LNG) Facilities within the United States. This guideline is intended to provide consistent wind load design recommendations for LNG facilities in the United States. LNG Facility design and approval to construct is regulated by the Federal Energy Regulatory Commission (FERC).

Wind load design for LNG Facilities in the United States is governed by 49 CFR 193.2067, which is maintained by the Department of Transportation (DOT). Requirements set forth in the Code of Federal Regulation (CFR) take precedence over requirements defined in NFPA 59A, Standard for the Production, Storage and Handling of Liquefied Natural Gas (LNG), and *ASCE 7*. Typically, the CFR requirements are more onerous than *ASCE 7* requirements. However, the engineer will need to verify which requirements control for the specific facility location. For locations outside of the United States, local building code requirements are to be satisfied at a minimum.

As part of the FERC permit process, the engineer must submit the LNG facility wind load design basis for review and approval by FERC and DOT to assure alignment with the government bodies. Failure to do so may incur delay in receipt of the Approval for Construction (AFC) permit.

4.5.2 Code of Federal Regulations

Requirements as of May 2007 of 49 CFR 193.2067 are listed below:

> For facilities with LNG containers of more than 70,000 gallons, the LNG Facility must maintain **structural and functional integrity** for:
>
> 1. An assumed **sustained wind velocity of not less than 150 miles per hour**, unless the Administrator finds a lower velocity is justified by adequate supportive data; or
>
> 2. The most critical combination of wind velocity and duration, with respect to the effect on the structure, having a **probability of exceeding a 50-year period of 0.5 percent or less**, if adequate wind data are available and the probabilistic methodology is reliable.

Even though the Code of Federal Regulations state requirements, interpretation in the industry is varied. This guideline is meant to provide a consistent interpretation of 49 CFR 193.2067 requirements. The committee contacted DOT to obtain alignment on the guidelines provided below. However, DOT was unable to provide comment prior to publication. Since DOT has not endorsed the below interpretation, the permit submittal must clearly define the wind design criteria for the entire facility and receive final acceptance by FERC and DOT.

4.5.3 Determination of Wind Speed

The lesser of 1) and 2) in Section 4.5.2 should be utilized in the design. However the application of the above wind speed should not be less than *ASCE 7* or other local jurisdictional requirements.

A sustained wind velocity of 150 miles per hour converts to 183 miles per hour 3-second gust. Sustained is a 1-minute average according to the National Weather Bureau. The conversion is performed utilizing the Durst Curve from the Chapter 6 Commentary of *ASCE 7*. This curve is shown in Figure 4.5.3-1.

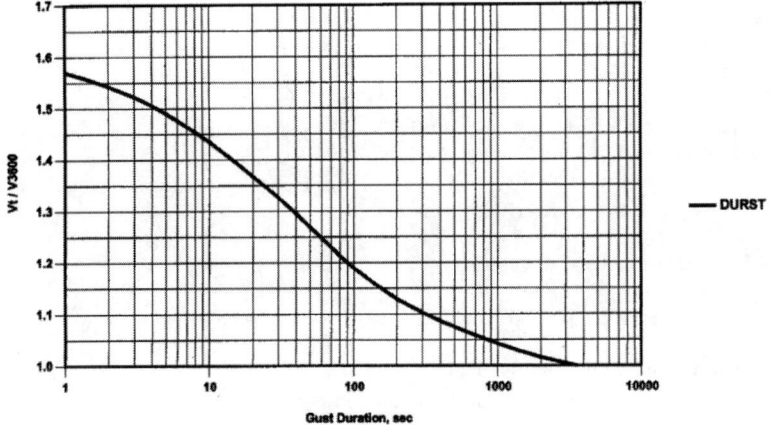

Figure 4.5.3-1: Maximum Speed Averaged over t_s to Hourly Mean Speed (ASCE 7-05 Figure C6-4)

Wind speed with a probability of exceeding 0.5 percent or less in a 50 year period equates to a 10,000-year mean recurrence interval (MRI). A wind speed map having such a return period was produced for ICC 500 standard (ICC, 2008) by Applied Research Associates, using the same methodology as the *ASCE 7* wind speed map. The 10,000-year MRI map is shown below in Figure 4.5.3-2. Wind speeds are peak 3-second gusts at 33 feet height over flat, open (Exposure C) terrain.

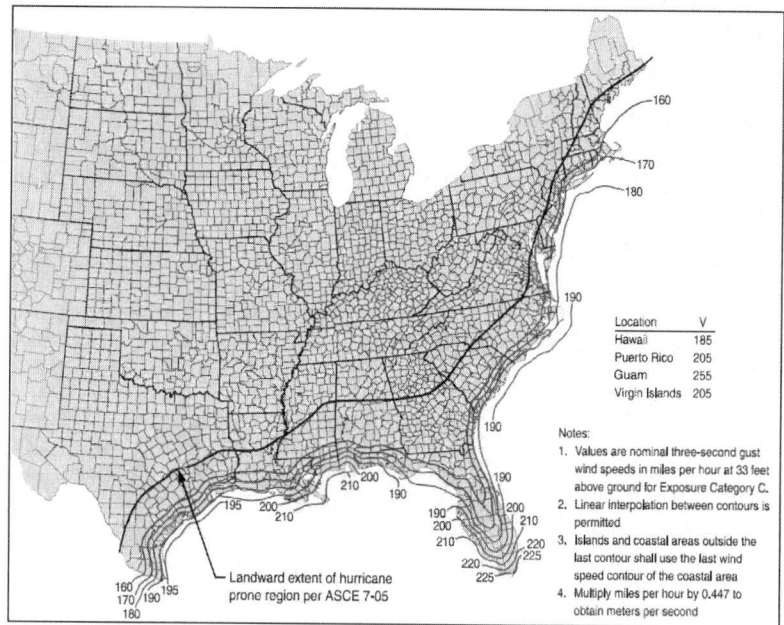

Figure 4.5.3-2: 10,000-year Mean Recurrence Interval Wind Speed (Source: ICC/NSSA Standard for the Design and Construction of Storm Shelters, ICC-500. Copyright 2008, International Code Council, Washington, D.C. Reproduced with permission. All rights reserved. www.iccsafe.org)

For coastal facilities in hurricane prone regions, the 10,000-year recurrence interval will likely be greater than the 150-mph sustained wind speed. For inland facilities that are not subject to hurricane effects, the 10,000-year recurrence interval will likely be less than the 150-mph sustained velocity requirement.

4.5.4 Wind Speed Applicability

The "LNG Facility" for which the CFR wind design requirements apply includes the LNG container, the impounding system, system or components required to isolate the LNG container and the fire protection system, whose failure could affect the integrity of the LNG container. The balance of the facility would be designed in accordance with *ASCE 7* wind load requirements, including use of the 3 second gust wind speeds that are appropriate for the facility location. The balance of the facility would include, but not be limited to, the following items: buildings, vaporizers, HP send out pumps, pipe racks, etc. The Engineer should consult with the Owner for agreement with the above philosophy prior to implementation.

Comment: Wind speeds listed in the CFR are for upper Category 4 Hurricanes, which are reasonably predictable events. Current operating practices in Petrochemical and LNG facilities are to shut down the facility to a minimal level prior to the hurricane making landfall. Since the facility is not "operating", the intent would be to maintain safe containment of the LNG. The "LNG Facility" would be designed to maintain structural and functional integrity. However, the balance of the facilities may suffer some level of minimal damage after such an event. The damage suffered by the balance of the facility would not compromise the impoundment of the LNG.

4.5.5 Importance Factor

An Importance Factor of 1.0 should be applied to the wind velocities listed in the CFR for the "LNG Facility". For the balance of the facility, use an Importance Factor per ASCE 7 requirements.

Comment: Importance Factors defined by *ASCE 7* are utilized to address the reliability of the structure by adjusting the recurrence interval. *ASCE 7* wind maps are based on a 50 year mean recurrence interval wind velocity. Since the CFR already dictates the recurrence interval, it is not necessary to apply an importance factor greater than 1.0.

4.5.6 Load Factors

For the "LNG Facility" and the balance of the facility, the application of *ASCE 7* load factors for Load Resistance Factor Design and Strength Design should be applied.

Comment: In strength design, buildings and structures are designed to resist the factored loads so that appropriate strength limitations are not exceeded. Load Factors are meant to address deviations in actual load from nominal load, uncertainties in the analysis, and the probability that one or more extreme events will occur simultaneously. Typically a load factor of 1.6 is applied to the wind load.

4.6 Evaluation of Wind Loads on Existing Structures

4.6.1 Introduction

In a petrochemical facility, it is common for process structures and pipe racks to be altered or modified from their original installation. Proposed changes and modifications to an existing facility must be evaluated for their impact on existing equipment support structures and foundations.

The engineering evaluation of existing structures and foundations should generally be approached in a different manner than the design of new facilities. Conservative design loads and a conservative application of the design loads are common practices in the design of new facilities. However, this same approach may not be warranted when evaluating existing structures. Conservatism in new design generally has minimal impact on the installed cost of the structure, but the same conservatism in the evaluation of an existing structure can have a significant negative cost and schedule impact on a project and may result in unnecessary upgrades.

The following sections provide references and summaries of current industry guidelines for the evaluation of existing structures and foundations.

4.6.2 References to Industry Codes and Guidelines

Several model building codes and industry guidelines address additions and alterations to existing buildings and structures. The following references provide some guidance regarding the evaluation and application of wind loading on modified structures.

The following references use either a 5% guideline (generally for gravity loads) or a 10% guideline (for lateral loads) for determining when upgrades of structural elements are required. When the modification does not increase the force (or increase the demand to capacity ratio) in an existing structural element by more than 5% or 10%, whichever is applicable, then a structural upgrade of that element is not required.

The engineer is referred to Section 4.6.4.2 in this document for a caution regarding the need to consider any past modifications (or structural upgrades) when evaluating the impact of a new modification.

Process Industry Practices (PIP) STC01015 Structural Design Criteria, September 2007

Section 4.4 states:

> "If the owner and engineer of record agree that the integrity of the existing structure is 100% of the original capacity based on the design code in effect at the time of the original design, structural designs should be performed in accordance with the following:
>
> 4.4.1 If additions or alterations to an existing structure do not increase the force in any structural element or connection by more than 5%, no further analysis is required.
>
> 4.4.2 If the increased forces on the element or connection are greater than 5%, the element or connection shall be analyzed to show that it is in compliance with the applicable design code for new construction.
>
> 4.4.3 The strength of any structural element or connection shall not be decreased to less than that required by the applicable design code or standard for new construction for the structure in question."

International Building Code 2006 (IBC 2006)

Modifications (alteration, repair or addition) to existing buildings are addressed in Chapter 34 Existing Structures. Section 3403.2 states:

> "Additions or alterations to an existing structure shall not increase the force in any structural element by more than 5 percent, unless the increased forces on the element are still in compliance with the code for new structures, nor shall the strength of any structural element be decreased to less than that required by this code for new structures."

International Building Code 2009 (IBC 2009)

Additions to existing buildings are addressed in Chapter 34 Section 3403, and existing structural elements are addressed in Sections 3403.3 and 3403.4. The provisions in Section 3403.3 for members carrying gravity load are the same as the IBC 2006. However, the provisions in Section 3403.4 are new and address members carrying lateral load. Section 3403.4 states:

> "Any existing lateral load-carrying structural element whose demand-capacity ratio with the addition considered is no more than 10 percent greater than its demand-capacity ratio with the addition ignored shall be permitted to remain unaltered."

International Existing Building Code 2006 (IEBC 2006)

The IEBC 2006 also used the 5% guideline that was recommended in the IBC 2006. For purposes of work classification, changes to existing buildings are classified as:

1. Repairs. Repairs include patching, restoration or replacement of damaged materials. Buildings that have sustained substantial structural damage to the vertical elements of the lateral-force-resisting system shall be evaluated. The wind design level for repair shall be as required by the building code in effect at the time of the original construction, unless the damage was caused by wind, in which case the design level shall be as required by the code in effect at the time of the original construction or as required by IBC, whichever is greater.

2. Alterations. Alterations do not generally affect wind forces unless there are additions or eliminations of doors and windows that change the closed/open classification of the building and thus affect the wind forces on the building.

3. Additions. Additions to existing buildings are new construction and are required to comply with the IBC. Any elements of the existing lateral-force-resisting system subjected to increased lateral or vertical loads due to vertical or horizontal additions are required to comply with the IBC.

WIND LOADS FOR PETROCHEMICAL AND OTHER INDUSTRIAL FACILITIES 61

International Existing Building Code 2009 (IEBC 2009)

Additions to existing buildings are addressed in Chapter 10, and existing structural elements are addressed in Sections 1003.2 and 1003.3. The provisions in Section 1003.2 for members carrying gravity load are the same as the IEBC 2006. However, the provisions in Section 1003.3 are new and address members carrying lateral load. Subsections 1003.3.1 and 1003.3.2 require existing members subject to an increase in load to comply with the IBC wind provisions, but there are exceptions noted. For example, "Exceptions: 2. In other existing buildings where the lateral-force story shear is not increased by more than 10 percent cumulative."

4.6.3 Considerations for Wind Exposure Category

The design of new petrochemical structures is generally based on wind exposure Category C (representative of flat open country and grasslands). However, many existing pipe racks and equipment structures are located in areas of the plant that are more representative of exposure Category B.

For the evaluation of existing structures and foundations, a less conservative wind exposure category may be considered. The exposure category for a particular location should be established in accordance with *ASCE 7*, and it should be noted that *ASCE 7* permits interpolation between exposure categories (Reference section 6.5.6.3 Exposure Categories).

For each wind direction considered, an exposure category that adequately reflects the characteristics of ground roughness and surface irregularities should be determined for the site at which the structure is located. For some structures, it would be appropriate to use a different exposure category for different wind directions.

Caution: Care should be exercised when establishing the exposure category for structures near the plant boundary. Generally, the area outside the fence line should be considered as characteristic of exposure Category C.

The following items highlight additional considerations when considering the wind exposure category to be used when evaluating existing structures.

Surface Roughness B is defined as urban and suburban terrain with numerous closely spaced obstructions having the size of single family dwellings. To qualify as exposure Category B, a distance of 2600 feet or 20 times the maximum height of the structure shall prevail in the upwind direction, whichever is greater. There is also an exception when the mean height is less than or equal to 30 feet, and in this case the upwind distance may be reduced to 1,500 ft (Reference section 6.5.6.3).

Most petrochemical facilities are comprised of numerous, closely spaced vessels and structures, but usually have many structures in excess of 30 feet in height. The use of

Category B should be considered when the location relative to upwind ground roughness would allow. Figure 4.6.1 is an aerial photo of a petrochemical facility and represents the typical plant layout with closely spaced vessels, pipe racks, storage tanks and equipment structures.

For comparison purposes, the wind pressure for an exposure Category B is approximately 70 percent of the pressure for an exposure Category C at 30 feet elevation.

Figure 4.6.1. Aerial Photo of Typical Petrochemical Facility (Courtesy of FEMA)

4.6.4 Other Cautions and Considerations

4.6.4.1 Review of Design Basis

Some plant sites are under the jurisdiction of local building officials. For these locations, the basis for the engineering evaluation should be reviewed with and meet the requirements of the local building officials. For all sites, the basis for the engineering evaluation should always be reviewed up front with the Owner of the facility.

4.6.4.2 Potential for Past Modifications

Existing structures may have been modified since their original installation. Over the years, additional equipment may have been added to a structure, or the original structure may have been expanded horizontally or vertically. An overall gravity, wind or seismic evaluation of the structure may not have been performed at the time of a past modification. A review of the existing drawings can provide valuable information regarding past modifications and also provide information regarding any structural upgrades that were made at the time of the additions or modifications.

It is also possible that the original structure has been modified and potentially weakened, either by the removal of horizontal or vertical bracing members, or by the modification of individual beams or columns. Figure 4.6.2 is a photograph that captures an unusual field modification to an existing steel column in a structure.

Figure 4.6.2 Photo of a Field Modification to an Existing Steel Column

The available engineering drawings may not accurately represent actual field conditions. A site visit to the plant should be conducted to verify the as-built condition of all critical members in the existing structure, especially those that are directly impacted by the proposed modification.

4.6.4.3 Potential for Structural Deterioration

Existing structures in the plants are subject to harsh environments and may have experienced some deterioration and loss of strength. A site visit should be made to assess the current condition of an existing steel or concrete structure. During the site visit, the engineer should be on the lookout for signs of active corrosion, especially for signs of corrosion that may be somewhat hidden under fireproofing. Figure 4.6.3 is a photograph of typical cracking and deterioration of a concrete column which indicates underlying corrosion of the reinforcing steel.

64   WIND LOADS FOR PETROCHEMICAL AND OTHER INDUSTRIAL FACILITIES

Figure 4.6.3 Photo of Deterioration of a Concrete Column

A visual only assessment may not be adequate to assess the condition of the existing structure, and it may be necessary to conduct a formal condition survey of any questionable members. There are firms that specialize in conducting condition surveys, and their services should be utilized as appropriate.

4.7 Wind Load Analysis Uncertainty

Many of the structures that appear in petrochemical and other industrial facilities have unique and complex geometries that have received relatively little attention from wind engineering researchers. As such, it would seem that a designer's confidence in the accuracy of the available methods for estimating wind loads on these types of structures would be somewhat lower than for structures typically appearing in commercial or residential settings.

Amoroso and Levitan (2009a and 2009b) compiled available wind tunnel test results for model open frame structures and vertical vessels in order to assess the uncertainty of the wind load analysis methods available for these structure types. The values of the predicted force coefficients were compared to the measured results for each of 48 different model configurations. The predicted coefficients were calculated using the provisions from Chapter 4 of the previous version of this guide (ASCE, 1997) and also using the provisions contained in Chapter 5 of this updated version.

The predictions using both the original and updated provisions were found to slightly overestimate the wind loads for the model structures. The standard deviations of the ratios of the measured values to the predicted values were found to be comparable to the variability associated with net pressure coefficient estimation for typical enclosed building structures (Ellingwood and Tekie, 1999). The standard deviation for the estimates using the updated guide provisions were nearly one-third lower than for the estimates using the original guide provisions, indicating a reduction in overall uncertainty upon application of the updated provisions.

Although the results of this analysis ultimately showed that the probability of failure for a petrochemical structure under the action of wind may be slightly higher than for a regular building structure, the differences were indeed small. Furthermore, the main reason for the difference in overall reliability between the modeled petrochemical structures and regular buildings was that the net pressure coefficient estimates for regular buildings are systematically overestimated to a greater degree.

It should be noted that the findings reported in this section were based on wind tunnel models of comparatively simple open frame structures and vessels. The dataset did not include any very large and/or highly complex structures.

CHAPTER 5
RECOMMENDED GUIDELINES PART II:
ANALYTICAL DETERMINATION OF WIND LOADS

This committee has updated the 1997 recommended guidelines and commentary to reflect recent research findings and updates to *ASCE 7*. There is also guidance provided for additional equipment types. When used herein, *ASCE 7* refers to *ASCE/SEI 7-05*.

GUIDELINES	COMMENTARY
5.0 General	**C5.0 General**
Design wind forces for the main wind force resisting system and components should be determined by the equation $$F = q_z \, G \, C_f \, A \quad (\text{Eq. 5.1})$$ (where F is the applied wind force) using the following procedure:	The basic equation for design wind loading (Equation 5.1) is adopted from *ASCE 7* procedures for "Design Wind Loads on Other Structures" (*ASCE 7* Section 6.5.15, Eq. 6-28). The provisions of Chapter 5 of this report primarily provide guidance in selecting appropriate force coefficients and projected areas.
The velocity pressure, q_z, is determined in accordance with the provisions of *ASCE 7* Section 6.5.10.	
The gust effect factor, G (or G_f), is determined in accordance with the provisions of *ASCE 7*, Section 6.5.8.	G_f is used in place of G for flexible structures, defined by *ASCE 7* as structures with a fundamental frequency f<1 Hz. If the height divided by least horizontal dimension is greater than 4, a frequency check may be warranted.
	Procedures and guidance for computing gust effect factors are provided in the *ASCE 7* Section 6.5.8 and commentary.

Force coefficients C_f and corresponding projected areas A_f or A_e are determined from the provisions of Section 5 for structures and equipment.

5.1 Pipe Racks

Wind on the pipe rack should be calculated as outlined below. The provisions below apply to transverse wind loads on the pipe rack.

C5.1 Pipe Racks

It is the experience of this committee that pipe friction and anchor loads govern the longitudinal design loads of a typical pipe rack. However, the engineer should exercise judgment to ensure longitudinal wind loads are considered for unusual pipe rack configurations.

5.1.1 Tributary Area for Piping

The tributary area for piping should be based on the diameter of the largest pipe, D, plus 10% of the width of the pipe rack, W. This result is multiplied by the length of the pipes (bent spacing, L) to determine the tributary area.

D = largest pipe diameter in bent (including insulation)
W = width of the bent
L = bent spacing
A = tributary area
A = L (D + 10%W) (Eq. 5.2)

C5.1.1 Tributary Area for Piping

This area is based on the assumption that the wind will strike at an angle plus or minus from the horizontal with a slope of 1 to 10 and that the largest pipe is on the windward side. This corresponds to an angle of ± 5.7 degrees. In some cases the pipe rack longitudinal strut or stringer might fall in the shielding envelope and should be deleted from wind load considerations.

This is a reasoned approach that accounts for wind on all the pipes (or cable trays) and shielding of the leeward pipes (or cable trays). The basis for the selection was a review of existing practices prior to publication of the previous edition of this guide.

5.1.2 Tributary Area for Cable Trays

The tributary area for cable trays should be based on the height of the largest tray, h plus 10% of the width of the pipe rack, W. This result is multiplied by the length of the trays (bent spacing, L) to determine the tributary area.

h = height of cable tray
W = width of bent
L = bent spacing
Ac = area of cable tray
Ac = L (h + 10%W)   (Eq. 5.3)

5.1.3 Force Coefficients for Structural Members

For all structural members C_f = 1.8, or alternatively C_f = 2.0 at and below the first level and C_f = 1.6 for members above the first level. No shielding shall be considered for the leeward column line except as noted in C5.1.1.

5.1.4 Force Coefficient for Pipes

The force coefficient C_f = 0.7 should be used as a minimum.

C5.1.2 Tributary Area for Cable Trays

See commentary C5.1.1.

C5.1.3 Force Coefficients for Structural Members

The C_f was determined with guidance from *ASCE 7* Figure 6-22 with consideration of typical solidities above and below the first level. For more than two column lines shielding may be considered.

C5.1.4 Force Coefficient for Pipes

The force coefficient C_f, for pipe is taken from *ASCE 7*, Figure 6-21 for a round shape, with an h/D = 25, $D\sqrt{q_z}$ > 2.5, and a moderately smooth surface; that is C_f = 0.7. If the largest pipe is insulated, then consider using a C_f for a rough pipe dependant on the roughness coefficient of the insulation (D'/D).

WIND LOADS FOR PETROCHEMICAL AND OTHER INDUSTRIAL FACILITIES 69

5.1.5 Force Coefficient for Cable Trays

For cable trays the force coefficient $C_f = 2.0$.

5.2 Open Frame Structures

5.2.1 General

This section covers wind loads on open frame or lattice structures, with or without equipment, piping, electrical items, stairs, ladders, handrail, etc.

Wind loads should be calculated in accordance with the general procedures and provisions of *ASCE 7* for wind loads on "Other Structures" with the exceptions as noted.

5.2.1.1 Main Wind Force Resisting System

1. Wind forces acting on the structural frame and appurtenances (ladders, handrails, stairs, etc.) should be computed in accordance with 5.2.2.

    A somewhat more exact and detailed method of calculating the wind forces on the structural frame and appurtenances is presented in Appendix 5A. This method allows the designer to estimate frame forces for a wider range of frame spacing ratios and solidity ratios. This method also allows the design-

C5.1.5 Force Coefficient for Cable Trays

The force coefficient C_f, for cable trays is taken from *ASCE 7*, Figure 6-21 for a square shape with the face normal to the wind and with a h/D = 25; that is $C_f = 2.0$.

C5.2.1.1 Main Wind Force Resisting System

1. The methods used to calculate wind loads on an open frame structure in 5.2 and Appendix 5A were adapted from theoretical work (Cook, 1990), wind tunnel testing results (Georgiou), and related analytical work (Willford and Allsop, 1990; Nadeem and Levitan, 1997) dealing with wind forces on identical, regularly spaced frames made of sharp-edged members.

    These methods have been extended to handle cases such as

er to calculate the wind angle producing the maximum force.

A simplified method that addresses high solidity structures (greater than 50% projected solidity including framing, equipment, and appurtenances) is presented in Appendix 5B.

frames of unequal solidity, the presence of secondary beams (beams not along column lines), and frames made up of rounded members (Willford and Allsop, 1990; Georgiou et al, 1981). Only limited experimental work has been carried out on some of these cases (Qiang, 1998; Qiang et al, 2004; and Amoroso and Levitan, 2009a). It is the opinion of the committee that it is not unreasonable to presume that for a structure which is not particularly unusual, irregular, or having too many appurtenances, the procedures of 5.2 and Appendix 5A should yield wind load results for the structure and the appurtenances together with reliability that is consistent with the other provisions in this chapter.

Research by Amoroso (2007) shows that the force coefficients for open frame structures with high projected solidity ratios can be reasonably estimated by considering a small number of key variables related to the overall geometry of the open frame structure. This approach does not require the designer to consider the combination of effects from framing, equipment, and other elements. This method is presented in Appendix 5B.

2. Wind forces on equipment, piping and cable trays located on or attached to the structure should be calculated according to the applicable provisions of

2. Scarce experimental work to date (Qiang, 1998; Qiang et al, 2004; and Amoroso and Levitan, 2009a) has considered the inclusion of three-dimensional

WIND LOADS FOR PETROCHEMICAL AND OTHER INDUSTRIAL FACILITIES 71

Chapter 5 and added to the wind forces acting on the frame in accordance with 5.2.6.

solidity (e.g., vessels, heat exchangers, etc.) placed in the framework. In general, it is expected that the total wind load on equipment will be less than the sum of the loads on the individual items due to shielding by the frame and other neighboring equipment.

Thus, the approach taken in 5.2.6 is the reduction of the total wind load on equipment by a multiplication factor η_{equip} to account for this shielding.

5.2.1.2 Force Coefficients for Components

Wind loads for the design of individual components, cladding, and appurtenances (excluding equipment, piping, and cable trays) should be calculated according to the provisions of *ASCE 7*. Based on common practice, force coefficients and areas for several items are given in Table 5.1. Force coefficients for open frame structures with partial cladding are given in 5.3.

5.2.2 Frame Load

For open frame structures, design wind forces for the main wind force resisting system should be determined by the equation:

$$F_S = q_z \, G \, C_f \, A_e \quad \text{(Eq. 5.1a)}$$

C5.2.2 Frame Load

The structure is idealized as two sets of orthogonal frames. The maximum wind force on each set of frames is calculated independently.

Note: In this equation, C_f accounts for the entire structure in the direction of the wind.

Table 5.1 Force Coefficients for Wind Loads on Components

Item	C_f	Projected Area
Handrail	2.0	0.80 sq. ft./ft.
Ladder without Cage	2.0	0.50 sq. ft./ft.
Ladder with Cage	2.0	0.75 sq. ft./ft.
Solid Rectangles & Flat Plates	2.0	
Round or Square Shapes	See *ASCE 7* Figure 6-21	
Stair w/ Handrail Side Elevation	2.0	Hand rail area plus channel depth
End Elevation	2.0	50 % of gross area

In Equation 5.1a, F_S is the wind force on structural frame and appurtenances, q_z and G are as defined in 5.0, and

1. The force coefficient C_f is determined from the provisions of 5.2.3.

2. The area of application of force A_e is determined per 5.2.5.

3. The design load cases are computed per 5.2.6.

5.2.2.1 Limitations of Analytical Procedure

Design wind forces are calculated for the structure as a whole.

The method is described for structures which are rectangular in plan and elevation.

C5.2.2.1 Limitations of Analytical Procedure

No information is provided about distribution of loads to individual frames. However, it should be noted that the windward frame will experience a larger percentage of the total wind force than any other frames, except possibly for the case where the solidity ratio of the windward frame is much less than that of other frames.

5.2.3 Force Coefficients

The force coefficient for a set of frames shall be calculated

C5.2.3 Force Coefficients

Force coefficients C_{Dg} are obtained from Figure 5.1 (see

by

$$C_f = C_{Dg} / \varepsilon \quad \text{(Eq. 5.2)}$$

where

C_{Dg} is the force coefficient for the set of frames given in Figure 5.1, and ε is the solidity ratio calculated in accordance with 5.2.4.

Alternately, C_f may be determined using Appendix 5A.

Force coefficients are defined for wind forces acting normal to the frames irrespective of the actual wind direction.

The frame spacing ratio is equal to S_F / B where S_F is the frame spacing in the direction parallel to the wind and B is the frame width as shown in Figure 5.1.

C5.2.1.1) or Appendix 5A. A single value is obtained for each axis of the structure. This value is the maximum force coefficient for the component of force acting normal to the frames for all horizontal wind angles. Although the wind direction is nominally considered as being normal to the set of frames under consideration, the maximum force coefficient occurs when the wind is not normal to the frames (see C5.2.6.1 and 5A.1). The angle at which the maximum force coefficient occurs varies with the dimensions of the structure, the solidity, number of frames, and frame spacing.

A method to estimate this angle is given in the Appendix 5A, which also provides C_f values for a larger range of S_F / B and ε values than Figure 5.1.

The force coefficients C_{Dg} were developed for use on the gross area (i.e., envelope area) of the structure as used by the British wind loading standard (Willford and Allsop, 1990). These are converted to force coefficients which are applied to solid areas as used in *ASCE 7* by Equation 5.2.

The force coefficients C_{Dg} were developed from wind tunnel tests for structures with a vertical aspect ratio (ratio of height to width perpendicular to the flow direction) of four.

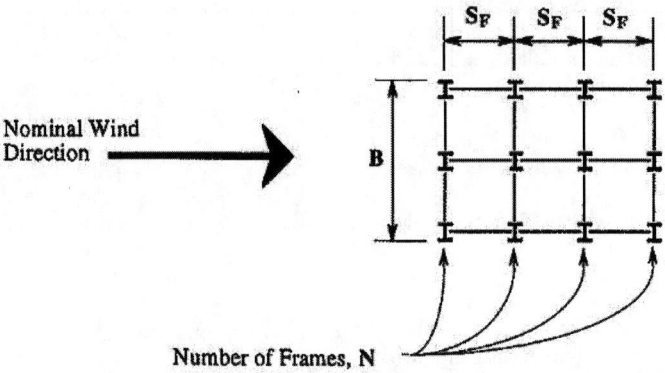

Figure 5.1 Force Coefficients, C_{Dg}, for Open Frame Structures

Notes:

(1) Frame spacing ratio is defined as S_F / B.

(2) Frame Spacing, S_F, is measured from centerline to centerline.

(3) Frame width, B, is measured from outside edge to outside edge.

(4) Number of frames, N, is the number of framing lines normal to the nominal wind direction (N = 4 as shown).

(5) Linear interpolation may be used for values of S_F / B not given on the following pages.

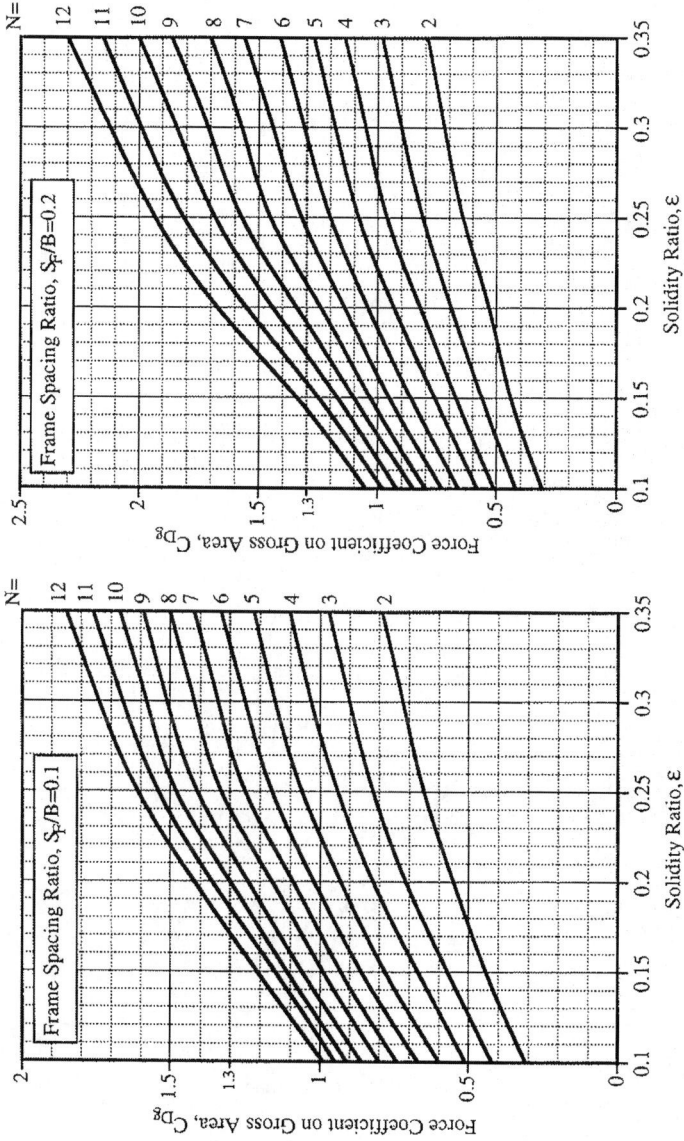

Figure 5.1   Force Coefficients, C_{Dg}, for Open Frame Structures (cont'd)

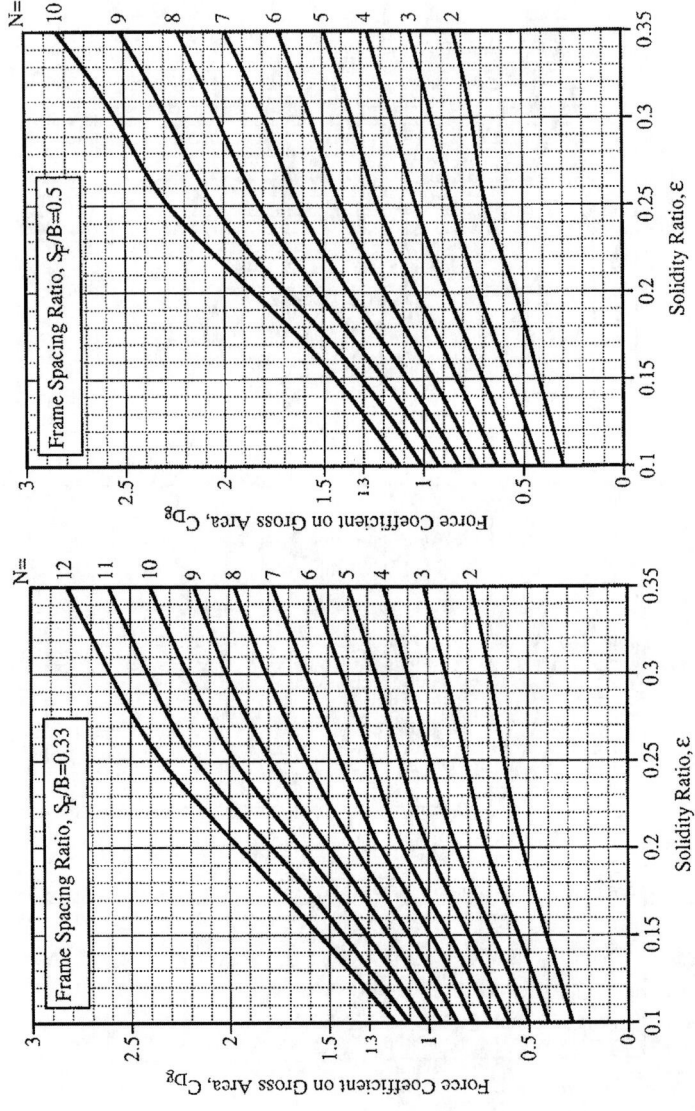

Figure 5.1 Force Coefficients, C_{Dg}, for Open Frame Structures (cont'd)

WIND LOADS FOR PETROCHEMICAL AND OTHER INDUSTRIAL FACILITIES 77

Although vertical aspect ratio does not play a large role in determining overall loads, the coefficients given in Figure 5.1 may be slightly conservative for relatively shorter structures and slightly unconservative for relatively taller structures.

Force coefficients C_{Dg} are applicable for frames consisting of typical sharp-edged steel shapes such as wide flange shapes, channels and angles. Reference Georgiou et al (1981) suggests a method to account for structures containing some members of circular or other cross sectional shape.

5.2.4 Solidity Ratio

The solidity ratio ε is given by:

$$e = A_S / A_g \quad \text{(Eq. 5.3)}$$

where A_g is the gross area (envelope area) of the windward frame and A_S is the effective solid area of the windward frame defined by the following:

5.2.4.1 The solid area of a frame is defined as the solid area of each element in the plane of the frame projected normal to the nominal wind direction. Elements considered as part of the solid area of a frame include beams, columns, bracing, cladding, stairs, ladders, handrails, etc. Items such as

C5.2.4 Solidity Ratio

Reference Willford and Allsop (1999) present a method to account for the effects of secondary floor beams (beams not in the plane of a frame). Use of this method may result in a small increase in the total wind force on the structure. With the associated uncertainties with the determination of the wind forces, this minor addition may be ignored.

5.2.4.2 The presence of flooring or decking does not cause an increase of the solid area of 5.2.4.1 beyond the inclusion of the thickness of the deck. Load–reducing effects of solid flooring (not grating) may be considered by adjusting the wind load by up to the shielding factor η_{floor}.

$$\eta_{floor} = 1 - 0.2\, (A_{fb} / A_S) \quad (5.4)$$

equipment, piping and cable trays are not included in calculation of the solid area of a frame; wind loads on these items are calculated separately.

where A_{fb} / A_S is the ratio of the projected area contributed by horizontal beams (in the vertical windward plane) supporting solid floors to the total projected solid area. This factor should be applied to the frame force coefficient.

5.2.4.3 For structures with frames of equal solidity, the effective solid area A_S should be taken as the solid area of the windward frame.

5.2.4.4 For structures where the solid area of the windward frame exceeds the solid area of the

C5.2.4.2 Recent experimental work (Qiang, 1998; Qiang et al, 2004; and Amoroso and Levitan, 2009a) indicates that the presence of solid decking decreases the wind forces compared to those of a bare frame. A history of research on bluff bodies with wake splitter plates reveals a reduction in drag due to a disruption of the vortex shedding pattern. This mechanism may explain the reductions in drag observed for open frames with solid floors. The relationship for the reduction factor given here is an empirical representation of the reduction of the wind load as a function of the solid area provided by floor beams (Amoroso and Levitan, 2009a).

No research related to open grating floors has been published. There is no evidence that open grating floors will significantly affect the wind forces on the structure.

C5.2.4.4 The force coefficients of Figure 5.1 were developed for sets of identical frames. Research

other frames, the effective solid area A_S should be taken as the solid area of the windward frame.

shows that the solidity of the windward frame is the most critical (Cook, 1990; Whitbread, 1980), leading to the recommendation. This provision is likely to yield slightly conservative loads, since the greater the solidity of the windward frame with respect to the other frames, the greater the shielding of the other frames.

5.2.4.5 For structures where the solid area of the windward frame is less than the solid area of the other frames, the effective solid area A_S should be taken as the average of all the frames.

5.2.4.6 When vertical bracing members in frames parallel to the nominal wind direction are present, the vertical projected area normal to the nominal wind direction for the vertical bracing members shall be added to A_S. Regardless of the configuration of the vertical bracing (diagonal bracing, chevron bracing, K-bracing, X-bracing, etc.) or arrangement (bracing located totally in one bay or distributed among several bays of a bent), the vertical projected area of only one brace member per story per braced bent parallel to the wind direction shall be considered.

C5.2.4.6 Recent experimental work (Amoroso and Levitan, 2009a) indicates that neglecting the contributions of vertical bracing to the solid area for wind directions nominally parallel to the plane of the bracing can lead to unconservative estimates of the wind load on an open frame structure.

5.2.5 Area of Application of Force

A_e shall be calculated in the same manner as the effective solid area in 5.2.4 except that it is for the portion of the structure height consistent with the velocity pressure q_z.

5.2.6 Design Load Cases

The total wind force acting on the structure in a given direction, F_T, is equal to the sum of the wind loads acting on the structure and appurtenances (F_S), plus the wind load on the equipment and vessels (per 5.4), plus the wind load on piping. See Figure 5.2 for complete definitions of F_T and F_S.

If piping arrangements are not known, the engineer may assume the piping area to be 10% of the gross area of the face of the structure for each principal axis. A force coefficient of 0.7 should be used for this piping area.

The two load cases shown in Figure 5.2 should be considered.

5.2.6.1
Frame load + equipment load + piping load (F_T) for one axis, acting simultaneously with 50% of the frame load (F_S) along the other axis, for each direction.

Combination of wind with other loads shall be computed in accordance with *ASCE 7*, Section 2.0.

C5.2.6 Design Load Cases

In some cases, this design load will exceed the load which would occur if the structure were fully clad. It is also possible that the wind load on just the frame itself (before equipment loads are added) will exceed the load on the fully clad structure. This happens most often for structures with at least 4 to 5 frames and relatively higher solidities. This phenomenon is very clearly demonstrated in Walshe (1965), which presents force coefficients on a building for 10 different stages of erection, from open frames to partially clad to the then fully clad building. The wind load on the model when fully clad is less than that during several stages of erection.

C5.2.6.1
While the maximum wind load normal to the frame for a structure consisting of a single frame occurs when the wind direction is normal to the plane of the frame, this is not the case for a structure with multiple planes of frames. The maximum load normal to the plane of the frames occurs when the wind direction is typically 10 to 45 degrees from the normal (Willford and Allsop, 1990). This is due to the

WIND LOADS FOR PETROCHEMICAL AND OTHER INDUSTRIAL FACILITIES 81

fact that for oblique winds there is no direct shielding of successive columns and a larger area of frame is therefore exposed to the wind directly (without shielding) as the wind angle increases. Thus, the maximum wind load on one set of frames occurs at an angle which will also induce significant loads on the other set of frames (Willford and Allsop, 1990; Georgiou et al, 1981).

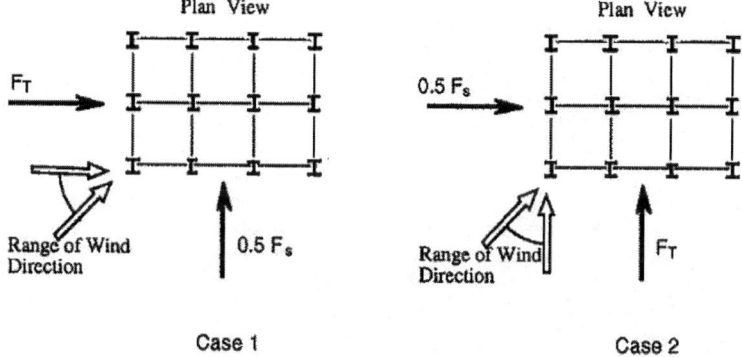

Figure 5.2 Design Load Cases

Notes:

(1) F_s denotes the wind force on the structural frame and appurtenances in the indicated direction (excludes wind load on equipment, piping, and cable trays.)

(2) F_T denotes the total wind force on the structure in the direction indicated, which is the sum of forces on the structural frame and appurtenances, equipment, and piping. If appropriate, the equipment load may be reduced by considering shielding effects per 5.2.6.2.

(3)   Load combination factors applied to F_s may alternatively be determined by the detailed method of Appendix 5A and used in place of the 0.5 values shown. These values shall be calculated separately for Case 1 and Case 2.

5.2.6.2 When, in the engineer's judgment, there is substantial shielding of equipment by the structure or other equipment on a given level in the wind direction under consideration, the wind load on equipment in 5.2.6.1 on that level may be reduced by the shielding factor, η_{equip}.

$$\eta_{equip} = \exp[-1.4\,(C_f\,\varepsilon)^{1.5}]$$

(Eq. 5.5)

The force coefficient, C_f, is for the frame according to 5.2.3 or Appendix 5A. The solidity ratio, ε, is defined in 5.2.4. This factor is applied to the equipment force coefficient.

The wind load on any equipment or portion thereof which extends above the top of the structure should not be reduced.

C5.2.6.2 Full and partial loading of structures given in *ASCE 7*, Section 6.5.12.3 was developed for clad structures only. The provisions of that paragraph are not applicable to open frame structures due to the different flow characteristics.

This provision is identical to a provision in the Australian wind load standard, AS/NZS 1170.2-2002, for reducing the wind load on cylindrical ancillaries located inside square-sectioned lattice towers. The provision in the Australian standard has its origin in data published by the Engineering Sciences Data Unit (ESDU) in the U.K. Recent experimental work (Amoroso and Levitan, 2009a) suggests that this provision is more accurate than the provision for equipment shielding given in the first edition of this guide.

5.2.6.3 Horizontal Torsion

Horizontal torsion (torsion about the vertical axis) may be a factor for open frame structures. The engineer should consider the possibility of torsion in the design.

C 5.2.6.3 Horizontal Torsion

The line of action of the wind load may not coincide with the center of rigidity of the structure. In this case, the wind force may produce torsional loads on the structure.

Consideration should be given to the application points of the wind load, especially in cases where the building framing is irregular and/or equipment locations are not symmetric.

5.3 Partially Clad Structures

This section is intended to address structures with cladding on less than four exterior walls. The wind loads on such structures vary considerably depending on the cladding arrangement and the wind direction. The forces calculated in this section are intended to be applied to the main wind force resisting system and are not intended for design of the components and cladding.

The different configurations of partially clad structures are illustrated in Figure 5.3. For all cases, forces along both axes should be applied simultaneously.

For open frames with only one vertical face having cladding $C_f = 1.4$ for forces acting normal to the clad face. For forces acting parallel to the clad face, C_f shall be determined using the methods for open frames in Section 5.2 or Appendix 5A.

For open frames with cladding on two opposite, parallel faces $C_f = 2.3$ for forces acting normal to the clad faces.

C5.3 Partially Clad Structures

The force coefficients provided in this section are derived from wind tunnel tests on partially clad structures (Amoroso et al, 2010). These experiments were of limited parametric extent, and therefore the guide provisions should not be interpreted as being all-inclusive. A significant and relevant general finding of the study was that considerably higher force coefficients can occur for partially clad arrangements than for fully clad structures of the same envelope geometry. Furthermore, high force coefficients are often generated simultaneously for both orthogonal structural axes. These findings are represented in the provisions of this section.

When determining wind loads on open frame structures, it is often advantageous to perform calculations for each of several individual stories along the structure height. This procedure takes advantage of the changes in the vertical velocity profile and accounts for changes in the geometric arrangement of the structure. Due to the high force coefficients associated with partially

For forces acting parallel to the clad faces, C_f shall be determined using the methods for open frames in Section 5.2 or Appendix 5A.

For open frames with cladding on two adjacent, perpendicular, vertical faces, C_f = 2.0 for forces along both structural axes when the unclad faces are positioned generally windward of the clad faces. When the clad faces are positioned windward of the unclad faces, C_f = 1.5 for forces acting along each structural axis.

For open frames with cladding on three vertical faces C_f = 1.5 for forces acting normal to the unclad face when the unclad face is positioned on the windward side of the structure. When the unclad face is positioned on the leeward side of the structure, C_f = 1.3 for forces acting normal to the unclad face. C_f = 1.3 for forces acting along the axis parallel to the unclad face.

clad structures, using this procedure is particularly important when the partial cladding does not extend over the full height of the structure.

WIND LOADS FOR PETROCHEMICAL AND OTHER INDUSTRIAL FACILITIES

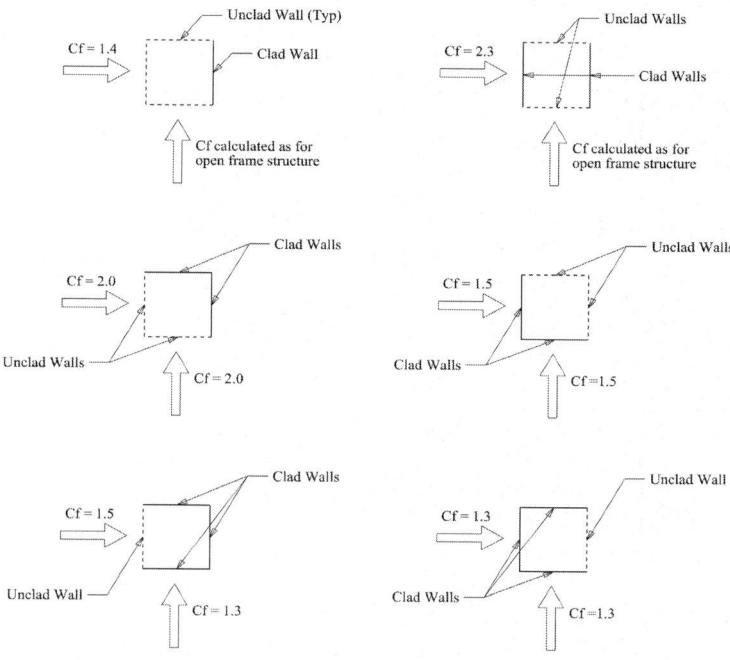

Figure 5.3-1 Configurations for Partially Clad Structures

5.4 Pressure Vessels

Where vessel and piping diameters are specified, it is intended that insulation, if present, be included in the projected area. Insulation should not be included for stiffness when checking H/D for dynamic characteristics.

C5.4 Pressure Vessels

For tall slender vessels, vortex shedding may cause significant oscillating force in the crosswind direction. This means that the structure may experience significant loads in both the along wind and crosswind directions at the same time. Crosswind forces such as vortex shedding are not addressed in this document.

5.4.1 Vertical Vessels

Use *ASCE 7* to calculate velocity pressures and to obtain the appropriate Gust Effect Factor, G_f, based on the governing empty or operating vessel frequency.

5.4.1.1 For those cases when fluid may be present inside the vessel during an extreme wind event, the designer should consider using the Detailed Method of *ASCE 7* to calculate the Gust Effect Factor, G_f and consider both cases when the vessel is empty as well as full of fluid. The detailed method of *ASCE 7* requires the engineer to determine the frequency of the vessel. The frequency can be determined by the following formulas:

$$f = 1/T \text{ (sec)} \quad \text{(Eq. 5.6)}$$

C5.4.1.1 The presence of fluid inside a vertical vessel could have a notable effect on the vessel frequency and, consequently, the total lateral load acting on the vessel due to wind. The Detailed Method of *ASCE 7* provides guidance to calculate the Gust Effect Factor, G_f. This procedure requires estimation of the vessel frequency, which can be accomplished using the equation for natural frequency stated.

A percentage of the empty weight of the vessel is usually added to the weight of the vessel to account for the weight of piping which is not included in the initial weight of the empty or operating weights. This percentage increase will affect the vessel frequency. When calculating this percentage increase, the engineer should be cautious to us an appropriate percentage increase based on the empty weight of the vessel compared to the additional weight of piping. A common percentage is usually around 10% of the empty weight of the vessel.

Natural Period of Vibration – Uniform Vertical Cylindrical Steel Vessel

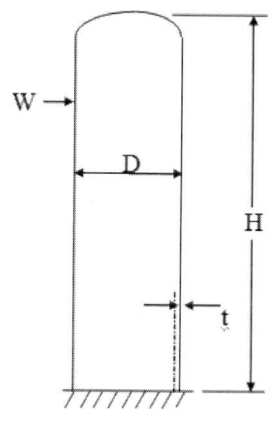

$$T = \frac{7.78}{10^6}\left(\frac{H}{D}\right)^2 \sqrt{\frac{12WD}{t}} \quad \text{(Eq. 5.7)}$$

where: T = Period (sec)
W = Weight (lb/ft)
H = Height (ft)
D = Diameter (ft)
t = Shell Thickness (inch)

Natural Period of Vibration – Non-Uniform Vertical Cylindrical Vessel

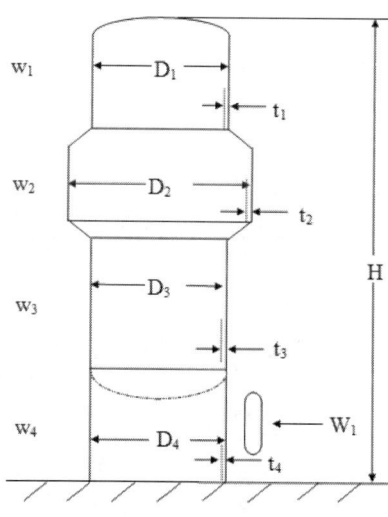

$$T = \left(\frac{H}{100}\right)^2 \sqrt{\frac{\sum w.\Delta\alpha + \frac{1}{H}\cdot\sum W.\beta}{\sum E.D^3.t.\Delta\gamma}}$$

(Eq. 5.8)

Where:

T = period (sec)
H = overall height (ft)
w = distributed weight (lbs/ft) of each section
W = Weight (lb) of each Concentrated Mass
D = diameter (ft) of each section
t = shell thickness (inch) of each section
E = modulus of elasticity (millions of psi)
α, β, and γ are coefficients for a given level depending on h_x/H ratio of the height of the level above grade to the overall height. $\Delta\alpha$ and $\Delta\gamma$ are the difference in the values of α and γ, from the top to the bottom of each section of uniform weight, diameter and thickness. β is determined and for each concentrated mass.

Values of α, β, and γ are tabulated on Table 5.2.

Table 5.2 Coefficients for Determining Period of Vibration of Free-Standing Cylindrical Shells with Non-Uniform Cross Section and Mass Distribution

h_x/H	α	β	γ	h_x/H	α	β	γ
1.00	2.103	8.347	1.000000	0.50	0.1094	0.9863	0.95573
0.99	2.021	8.121	1.000000	0.49	0.0998	0.9210	0.95143
0.98	1.941	7.898	1.000000	0.48	0.0909	0.8584	0.94683
0.97	1.863	7.678	1.000000	0.47	0.0826	0.7987	0.94189
0.96	1.787	7.461	1.000000	0.46	0.0749	0.7418	0.93661
0.95	1.714	7.248	0.999999	0.45	0.0578	0.6876	0.93097
0.94	1.642	7.037	0.999998	0.44	0.0612	0.6361	0.92495
0.93	1.573	6.830	0.999997	0.43	0.0551	0.5372	0.91854
0.92	1.506	6.626	0.999994	0.42	0.0494	0.5409	0.91173
0.91	1.440	6.425	0.999989	0.41	0.0442	0.4971	0.90443
0.90	1.377	6.227	0.999982	0.40	0.0395	0.4557	0.89679
0.89	1.316	6.032	0.999971	0.39	0.0351	0.4167	0.88864
0.88	1.256	5.840	0.999956	0.38	0.0311	0.3801	0.88001
0.87	1.199	5.652	0.999934	0.37	0.0275	0.3456	0.87033
0.86	1.143	5.467	0.999905	0.36	0.0242	0.3134	0.86123
0.85	1.090	5.285	0.999867	0.35	0.0212	0.2833	0.85105
0.84	1.038	5.106	0.999817	0.34	0.0185	0.2552	0.84032
0.83	0.938	4.930	0.999754	0.33	0.0161	0.2291	0.82901
0.82	0.939	4.758	0.999674	0.32	0.0140	0.2050	0.81710
0.81	0.892	4.589	0.999576	0.31	0.0120	0.1826	0.80459
0.80	0.847	4.424	0.999455	0.30	0.010293	0.16200	0.7914
0.79	0.804	4.261	0.999309	0.29	0.008769	0.14308	0.7776
0.78	0.762	4.102	0.999133	0.28	0.007426	0.12576	0.7632
0.77	0.722	3.946	0.998923	0.27	0.006249	0.10997	0.7480
0.76	0.683	3.794	0.998676	0.26	0.005222	0.09564	0.7321
0.75	0.646	3.645	0.998385	0.25	0.004332	0.08267	0.7155
0.74	0.610	3.499	0.998047	0.24	0.003564	0.07101	0.6981
0.73	0.576	3.356	0.997656	0.23	0.002907	0.06056	0.6800
0.72	0.543	3.217	0.997205	0.22	0.002349	0.05126	0.6610
0.71	0.512	3.081	0.996689	0.21	0.001878	0.04303	0.6413
0.70	0.481	2.949	0.996101	0.20	0.001485	0.03579	0.6207
0.69	0.453	2.820	0.995434	0.19	0.001159	0.02948	0.5902
0.68	0.425	2.694	0.994681	0.18	0.000893	0.02400	0.5769
0.67	0.399	2.571	0.993834	0.17	0.000677	0.01931	0.5536
0.66	0.374	2.452	0.992885	0.16	0.000504	0.01531	0.5295
0.65	0.3497	2.3365	0.99183	0.15	0.000368	0.01196	0.5044
0.64	0.3269	2.2240	0.99065	0.14	0.000263	0.00917	0.4783
0.63	0.3052	2.1148	0.98934	0.13	0.000183	0.00689	0.4512
0.62	0.2846	2.0089	0.98739	0.12	0.000124	0.00506	0.4231
0.61	0.2650	1.9062	0.98630	0.11	0.000081	0.00361	0.3940
0.60	0.2464	1.8068	0.98455	0.10	0.000051	0.00249	0.3639
0.59	0.2288	1.7107	0.98262	0.09	0.000030	0.00165	0.3327
0.58	0.2122	1.6177	0.98052	0.08	0.000017	0.00104	0.3003
0.57	0.1965	1.5279	0.97823	0.07	0.000009	0.00062	0.2669
0.56	0.1816	1.4413	0.97573	0.06	0.000004	0.00034	0.2323
0.55	0.1676	1.3579	0.97301	0.05	0.000002	0.00016	0.1965
0.54	1.1545	1.2775	0.97007	0.04	0.000001	0.00007	0.1597
0.53	0.1421	1.2002	0.96683	0.03	0.000000	0.00002	0.1216
0.52	0.1305	1.1259	0.96344	0.02	0.000000	0.00000	0.0823
0.51	0.1196	1.0547	0.95973	0.01	0.000000	0.00000	0.0418

5.4.1.2 Simplified Method

If detailed information (number of platforms, platform size, etc.) is unknown at the time of design of the foundation/piles, the following approach may be used:

1. For the projected width, add 5-ft (1.52 m) to the diameter of the vessel, or add 3-ft (0.91 m) plus the diameter of the largest pipe to the diameter of the vessel, whichever is greater. This will account for platforms, ladders, nozzles and piping below the top tangent line.

2. The vessel height should be increased one (1) vessel diameter to account for a large diameter pipe and platform attached above the top tangent, as is the case with most tower arrangements.

3. The increases in vessel height or diameter to account for wind on appurtenances should not be used in calculating the H/D ratio for force coefficients or flexibility.

4. The force coefficient (C_f) should be determined from *ASCE 7*, Figure 6-21, assuming a rough surface.

C5.4.1.2 Simplified Method

1. With limited information on the vessel and appurtenances, this simple approach gives reasonably consistent results.

2. This method is an approximation to alleviate the need for some rather tedious calculations based on gross assumptions.

3. As noted in 5.4 previously, insulation should not be included in the h/D calculations.

4. Based on wind tunnel studies conducted by Amoroso [2007], the force was underestimated using the simplified method from the previous edition of this guideline. Therefore, a rough surface should be assumed which, in the absence of more detailed design information, will likely result in a more appropriate (i.e., higher) estimate of the total wind load. If ribbed insulation will be used, then the D'/D should be calculated.

5.4.1.3 Detailed Method

If most design detail items (platforms, piping, ladders, etc.) of the vessel are known, the following method should be used:

1. For the projected width, add 1.5-ft (0.46 m) to the vessel diameter to account for ladders, nozzles and piping 8-in (0.2 m) or smaller and add the diameter of the largest line coming from the top portion of the vessel.

2. The force coefficient (C_f) should be taken from *ASCE 7*, Figure 6-21 based on appropriate roughness at vessel surface.

3. For pipes outside the projected width of the vessel (defined in 1) larger than 8-in (0.2 m), including insulation, use the projected area of the pipe and use a force coefficient (C_f) of 0.7.

   For pipes inside the projected width of the vessel (defined in 1) larger than 8-in (0.2 m), including insulation, and more than 5 pipe diameters from the vessel surface, add the projected area of the pipe and use a force coefficient (C_f) of 0.7.

4. For platforms, use the projected area of the support

C5.4.1.3 Detailed Method

This method will provide more accurate values for foundation design.

3. C_f is determined from *ASCE 7* Figure 6-21 for a moderately smooth surface.

4. The front and back systems of railings are far enough apart to preclude shielding.

steel and a force coefficient (C_f) of 2.0.

For handrails, use the values for area and force coefficient from Table 5.1.

Where the railing extends beyond the vessel, the projected area of two (2) sets of railing systems for the projection beyond the vessel should be used.

5.4.1.4 When a vertical vessel is located at a center-to-center distance of three (3.0) diameters or less (based on the smaller of the two vessels) from a nearby vertical vessel, an increase of 20 percent is to be applied to the force coefficient value in the absence of model wind tunnel testing or full scale data. This increase is to be applied for wind directions oriented approximately perpendicular to the axis connecting the center lines of the vessels.

C5.4.1.4 Research has shown that wind loads for closely spaced cylinders and similar structures can be amplified due to aerodynamic interference effects. This provision and the provision immediately following are modeled after a similar provision in ASME STS-1-2006 for the wind loading of steel stacks.

The vessel manufacturer should be advised of the increased wind load for closely spaced vessels.

5.4.1.5 When a large diameter pipe (greater than 8-in.) is located at center-to-center distance of three (3.0) pipe diameters from the surface of a vertical vessel, the force coefficient applied to the pipe is to be increased by 20 percent in the absence of model wind tunnel testing or full-scale data. This increase is to be applied for wind directions oriented approximately perpendicular to the axis connecting the center lines of the pipe and the vessel.

5.4.2 Horizontal Vessels

5.4.2.1 No check for dynamic properties is required.

5.4.2.2 For the projected diameter, add 1.5 ft (0.46 m) to the insulated diameter to account for ladders, nozzles and pipe 8 in (0.2 m) (including insulation) or smaller.

5.4.2.3 For wind perpendicular to the long axis of the vessel (transverse wind), the force coefficient (C_f) should be determined from *ASCE 7*, Figure 6-21.

C5.4.2.3 Use B/D (Length/Diameter) to determine C_f.

5.4.2.4 For wind in the longitudinal direction, use C_f of 0.5 for a rounded head and 1.2 for a flat head.

C5.4.2.4 This wind direction will seldom control design of foundations.

5.4.2.5 For pipe larger than 8 in (0.2 m), including insulation, use the projected area of the pipe and use a force coefficient (C_f) of 0.7.

5.4.2.6 For platforms, use the projected area of the support steel and a force coefficient (C_f) of 2.0.

C5.4.2.6 The reason for projecting the front and back railing system is that they are far enough apart to preclude shielding.

For handrails, use the values for area and force coefficient from Table 5.1.

Use the projected area of each railing system.

5.4.2.7 For supports, use the actual projected area. C_f should be 1.3 for concrete pedestals. For steel supports, use the method

C5.4.2.7 The 1.3 factor is used because a pedestal is similar to a bluff rectangular body.

described for platforms.

5.4.3 Spheres

5.4.3.1 No check for dynamic properties is required.

5.4.3.2 For the projected diameter, add 1.5 ft (0.46 m) to the insulated diameter to account for ladders, nozzles and pipe 8 in (0.2 m) (including insulation) or smaller.

5.4.3.3 Use $C_f = 0.5$ (for vessel only). Supports should be evaluated separately.

C5.4.3.3 *ASCE 7* does not provided C_f values for spheres. Daughtery and Franzini (1977) provides drag coefficients for a sphere as a function of Reynolds Number (Re). For Re in the range of 10^5 the value is approximately equal to 0.5.

5.4.3.4 For pipe larger than 8 in (0.2 m), including insulation, use the projected area of the pipe and use a force coefficient (C_f) of 0.7.

C5.4.3.4 C_f is determined from *ASCE 7* Figure 6-21 for a moderately smooth surface.

5.4.3.5 For platforms, use the projected area for the support steel and a force coefficient (C_f) of 2.0.

For handrails, use the values for area and force coefficient from Table 5.1. Use the projected area of each railing system.

C5.4.3.5 The reason for projecting the front and back railing system is that they are far enough apart to preclude shielding.

5.4.3.6 For supports, use the actual projected area. C_f should be 1.3 for rectangular concrete columns and 0.7 for circular columns. For steel supports, use the method described for platforms.

5.5 Cooling Towers

Use *Method 1 – Simplified Procedure* of *ASCE 7* to calculate wind pressures on the MWFRS and on components and cladding.

If located on top of a building or other supporting structure in a manner where the top of the tower is at a height greater than 60 ft., then use *Method 2 – Analytical Procedure* of *ASCE 7*.

For the fan cylinder (or shroud), calculate wind pressures across its surface separately using the following equation:

$p_z = K_z K_{zt} (V/28)^2 I$ (Eq. 5.9)

[p_z in psf; V in mph]

$F_z = p_z H D$ (Eq. 5.10)

[F_z in lbs.]

H = height (ft.)
D = diameter (ft.)

C5.5 Cooling Towers

The Cooling Technology Institute (CTI) standard cites the *ASCE 7* standard as the design basis for calculating wind loads on wooden cooling towers.

The cylinder is similar in geometry and aspect ratio to open top tanks. The American Petroleum Institute (API) provides design specifications for such structures. The equation provided here was derived based on API and on *ASCE 7*.

WIND LOADS FOR PETROCHEMICAL AND OTHER INDUSTRIAL FACILITIES

5.6 Air Cooled Heat Exchanger (Air Coolers or Fin-Fans)

In Table 5.3, the force coefficients C_f for isolated solid blocks are provided to calculate wind loads on air coolers. The effective area A_e equals b times c.

C5.6 *ASCE 7* does not have force coefficients for solid blocks in open space. In the past, engineers used their own engineering judgment to calculate wind loads on air coolers.

Multiple cooler units can be treated as one larger body for wind load estimation purposes when the clear spacing between units in the along-wind direction is less than or equal to one-half the cross-wind dimension b of the individual unit. In this case the a dimension of the combined units is the sum of the a dimension for each unit plus the clear space.

This provision applies only to the wind load on the air cooler unit. The dimensions should not include platforms and other appurtenances outside the outline of the air cooler unit.

96 WIND LOADS FOR PETROCHEMICAL AND OTHER INDUSTRIAL FACILITIES

Table 5.3 C_f for Isolated Blocks in Datum Conditions

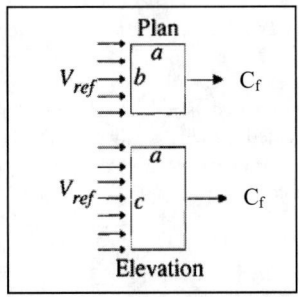

		a/b								Max C_f	
		0.00	0.25	0.50	1.00	1.50	2.00	2.50	3.00	C_f	a/b
	0.25	1.20	1.20	1.00	0.92	0.91	0.91	0.91	0.90	1.23	0.20
	0.50	1.16	1.21	1.16	0.97	0.92	0.91	0.91	0.90	1.21	0.34
	1.00	1.14	1.15	1.18	1.16	1.02	0.95	0.92	0.90	1.20	0.70
c/b	2.00	1.16	1.17	1.19	1.17	1.04	0.97	0.93	0.92	1.21	0.70
	5.00	1.21	1.22	1.24	1.22	1.11	1.03	0.99	0.97	1.27	0.70
	8.00	1.28	1.29	1.35	1.30	1.17	1.09	1.04	1.01	1.36	0.70
	10.00	1.31	1.34	1.41	1.36	1.24	1.14	1.08	1.04	1.46	0.70

Note: This Table is generated from Figure 1 of ESDU (1978) Fluid Forces, Pressures and Moments on Rectangular Blocks, ESDU Data Item 71016, Engineering Sciences Data Unit, London.

APPENDIX 5A
ALTERNATE METHOD FOR DETERMINING C_f AND LOAD COMBINATIONS FOR OPEN FRAME STRUCTURES

5A.1 Background

Maximum wind force normal to the face of a rectangular enclosed building occurs when the wind direction is normal to the building face. The same is true for wind load on a single frame or solid sign. However, this is not the case for an open frame structure with more than one frame. As the wind direction moves away from the normal and more toward a quartering wind, columns which once lined up neatly behind each other, shielding each other, become staggered and exposed to the full wind. Additionally, the area of the structure projected on a plane normal to the wind also increases.

The variation of the wind loads along each principal axis of a rectangular open frame structure with the direction of wind is shown in Figure 5.A.1, for the structure and wind angle of attack defined in Figure 5A.2. It can readily be seen that when one frame set experiences its maximum frame load "A" or "D", the frame set along the other axis experiences a wind force "C" or "B" respectively, thus the need for the load combinations of Section 5.2.6.1. In those provisions, the load at "C" is roughly estimated to be 50% of "A" and "B" is estimated to be 50% of the load at "D". In actuality, the loads on the secondary axis can range from about 25% to 75% of the primary axis load, depending on many factors including spacing ratio, number of frames, solidity ratio, etc. This appendix provides a method to obtain a better estimate of the simultaneously acting load on the secondary axis.

5A.2 Force Coefficients

This method provides force coefficients C_f for a greater range of ε and S_F/B values than the method of 5.2.3, as well as providing an estimate of α_{max}. For cases where both methods are applicable, they will generally yield very similar results. References *Nadeem* and *Nadeem / Levitan* discuss this method in greater detail. The procedure is as follows:

1. Determine ε, S_F/B, and N for the principal axis under consideration as per 5.2.4 and Figure 5.1.

2. Estimate the wind angle of attack which maximizes the force parallel to the axis under consideration.

$\alpha_{max} = (10 + 58\varepsilon)°$ for $3 \leq N \leq 5$ (Eq. 5A.1)
$\alpha_{max} = (16 + 52\varepsilon)°$ for $6 \leq N \leq 10$ (Eq. 5A.2)

3. Estimate the force coefficient C_f from Figure 5A.3 by the following procedure:

a. Determine C_f from Figure 5A.3(a) for wind angle of attack = α_{max} and appropriate spacing ratio S_F/B. This C_f is for a structure with N=3 frames and a solidity ratio of $\varepsilon = 0.1$.

b. Determine C_f from Figure 5A.3(b) for a structure with N=3 frames and a solidity ratio of $\varepsilon = 0.5$.

c. Interpolate between results of a) and b) for the actual solidity ratio, yielding a force coefficient for the correct spacing and solidity ratios, and N=3 frames, $C_{f,N=3}$.

d. Determine C_f from Figure 5A.3(c) for a structure with N=10 frames and a solidity ratio of $\varepsilon = 0.1$.

e. Determine C_f from Figure 5A.3(d) for a structure with N=10 frames and a solidity ratio of $\varepsilon = 0.5$.

f. Interpolate between results of d) and e) for the actual solidity ratio, yielding a force coefficient for the correct spacing and solidity ratios, and N=10 frames, $C_{f,N=10}$.

g. Determine C_f for the axis under consideration by interpolating between $C_{f,N=3}$ and $C_{f,N=10}$ for the actual number of frames.

Note that if the structure has exactly 3 or 10 frames, only steps (a-c) or (d-f) respectively need to be used. Similarly, if a structure has a solidity ratio very near to 0.1 or 0.5, only one interpolation between Figures 5A.3(a) and (c) or 5A.3(b) and (d) respectively would be necessary.

5A.3 Load Combinations

Section 5.2.6.1 specified the load combination of full wind load on the axis under consideration acting simultaneously with 50% of the frame wind load on the other axis. A more detailed method to estimate the wind load acting simultaneously on the secondary axis frames is given here.

1. Determine C_f for the principal axis under consideration as per 5.2.3 or 5A.2. If the provisions of 5.2.3 are used, α_{max} must still be determined as per 5A.2.

2. Determine the force coefficient C_f for the secondary axis from Figure 5A.3, using ε, S_F/B, and N values for the secondary axis and a wind angle of attack of (90° - α_{max}). Step 3 in Section 5A.2 explains how to obtain C_f from Figure 5A.3.

WIND LOADS FOR PETROCHEMICAL AND OTHER INDUSTRIAL FACILITIES   99

5A.4 Sample Calculations

The proposed method has been used to calculate the force coefficients for a structure whose plan is shown in Figure 5A.4. Summarizing the important frame set properties,

For winds nominally from west to east (i.e., winds normal to the N-S frame set)

ε = 0.136
N = 4
S_F = 25.4 ft. (7.75 m)
B = 68.9 ft. (21.0 m)
S_F/B = 25.4 / 68.9 = 0.369

For winds nominally from south to north (i.e., winds normal to the E-W frame set)

ε = 0.286
N = 5
S_F = 16.4 ft. (5.0 m)
B = 78.75 ft. (24.0 m)
SF/B = 16.4 / 78.75 = 0.208

Determining Force Coefficients: To determine the force coefficient for the E-W structural axis (winds nominally normal to the N-S frame), first estimate the wind angle of attack at which this maximum load will occur. Since $N = 4$ and $\varepsilon = 0.136$ for the N-S frame set, Equation 5A.1 yields

$$\alpha_{max} = 10 + 58 (0.136) = 18°$$

From Figures 5A.3(a) and (b), for $S_F/B = 0.369$, $C_f = 3.87$ and 2.10 for structures with $\varepsilon = 0.1$ and $\varepsilon = 0.5$, respectively, and $N = 3$ frames. Interpolating between these two values for $\varepsilon = 0.136$,

$$C_{f,N=3} = 3.87 - [(3.87 - 2.10) / 0.4] (0.136 - 0.1) = 3.71$$

From Figures 5A.3(c) and (d), for $S_F/B = 0.369$, $C_f = 10.08$ and 3.15 for structures with $\varepsilon = 0.1$ and $\varepsilon = 0.5$, respectively, and $N = 10$ frames. Interpolating between these two values for $\varepsilon = 0.136$,

$$C_{f,N=10} = 10.08 - [(10.08 - 3.15) / 0.4] (0.136 - 0.1) = 9.46$$

Interpolating between the two previous results of $C_{f,N=3} = 3.71$ and $C_{f,N=10} = 9.46$ for the case $N = 4$,

$$C_f = 3.71 + [(9.46 - 3.71) / 7] (4 - 3) = 4.53$$

gives the maximum force coefficient for the N-S set of frames, occurring near $\alpha_{max} = 18°$.

Determining Load Combinations: While the maximum wind load is acting on the N-S set of frames, the wind simultaneously acts on the E-W set of frames at angle of attack of $90° - 18° = 72°$.

The force coefficient for the E-W frames is determined as per step 2 of 5A.3.

For $\alpha = 72°$, $S_F / B = 0.208$ and $\epsilon = 0.286$, interpolation between Figures 5A.3(a) and (b) yields $C_{f,N=3} = 0.91$. Interpolation between Figures 5A.3(c) and (d) yields $C_{f,N=10} = 2.84$.

One more interpolation between $C_{f,N} = 3$ and $C_{f,N} = 10$ for $N = 5$ frames yields $C_f = 1.46$, which is the force coefficient for the wind load acting on the E-W frame set while the N-S set is experiencing its maximum wind load. This combination of loads is shown in Figure 5A.5(a).

This entire procedure should now be repeated assuming that maximum wind load acts on the E-W set of frames. For this case, $\alpha_{max} = 27°$, $C_f = 4.0$ for the E-W frames, with a simultaneously acting load of $C_f = 2.19$ for the N-S frame set, as shown in Figure 5A.5(b).

Note that for the case of full wind load on the E-W axis, use of this alternate procedure reduced the wind load acting simultaneously on the N-S axis from 50% (5.2.6.2) to $1.46/4.00 = 37\%$. The load combination for full wind on the N-S structure axis remained close to the recommended 50% at $2.19/4.53 = 48\%$.

WIND LOADS FOR PETROCHEMICAL AND OTHER INDUSTRIAL FACILITIES 101

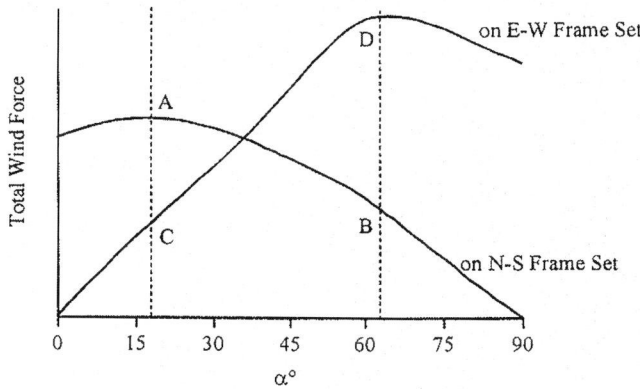

Figure 5A.1 Variation of Wind Load vs. Wind Direction

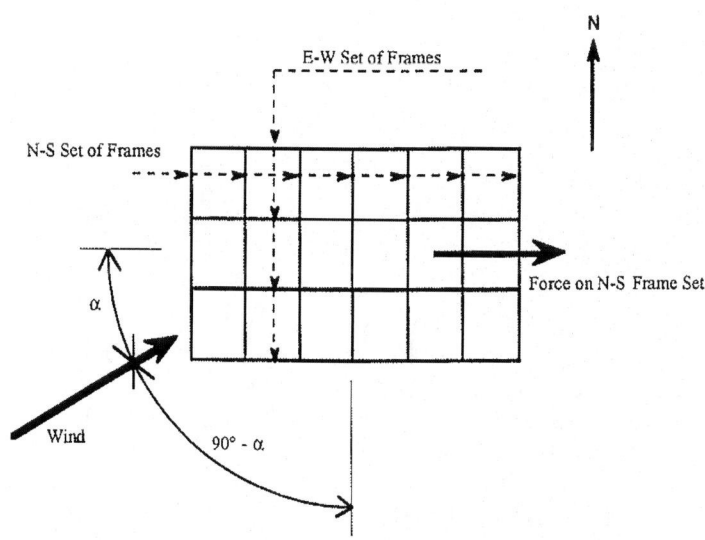

Figure 5A.2 Plan View of Structural Framing

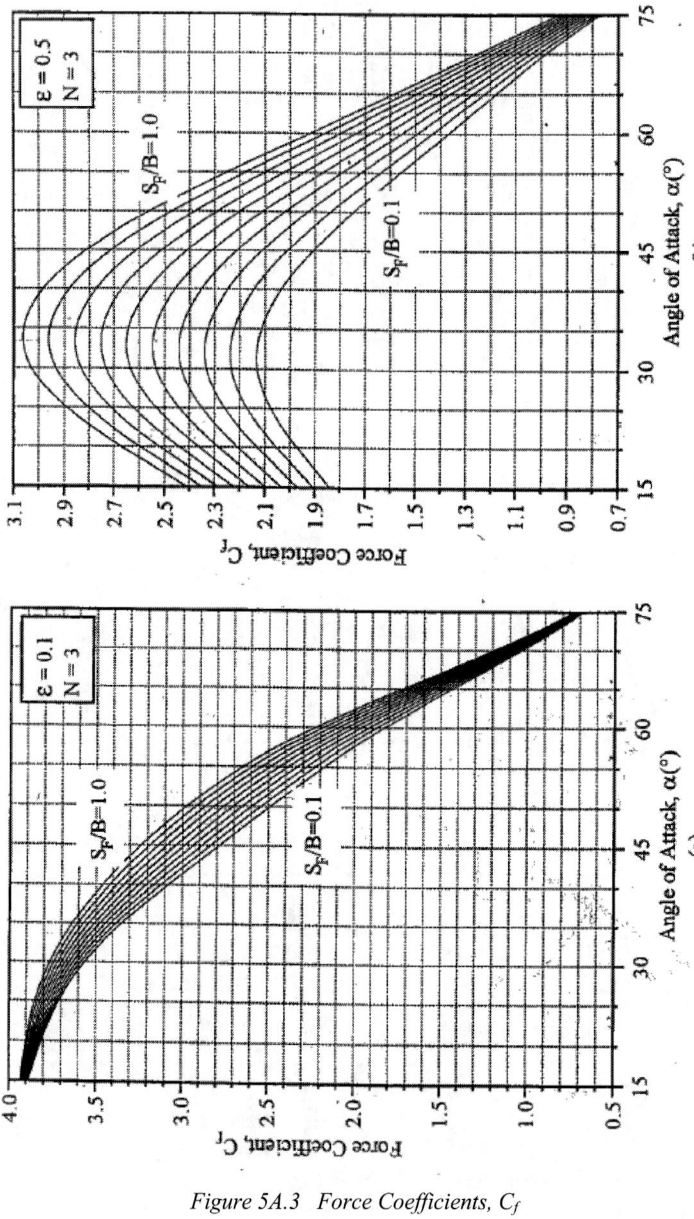

Figure 5A.3 Force Coefficients, C_f

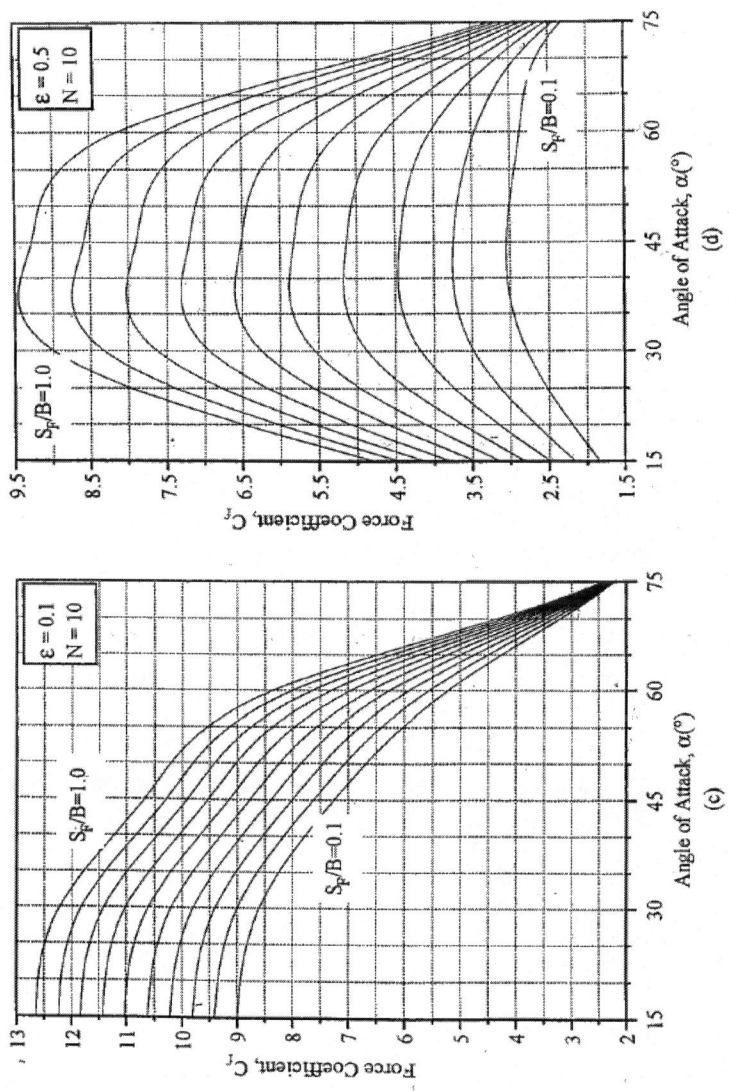

Figure 5A.3   Force Coefficients, C_f (cont'd)

104 WIND LOADS FOR PETROCHEMICAL AND OTHER INDUSTRIAL FACILITIES

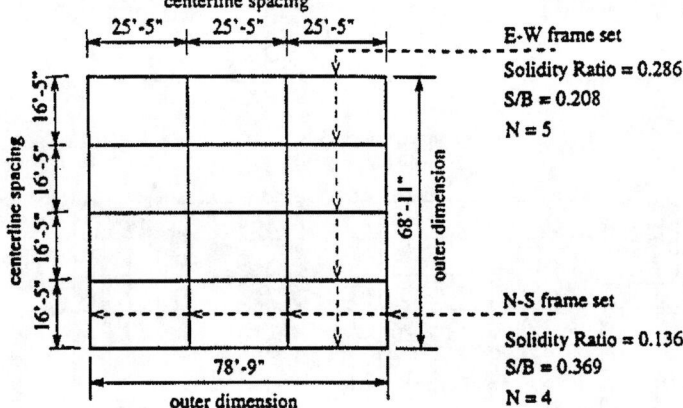

Figure 5A.4 Structural Framing Plan for Sample Calculations

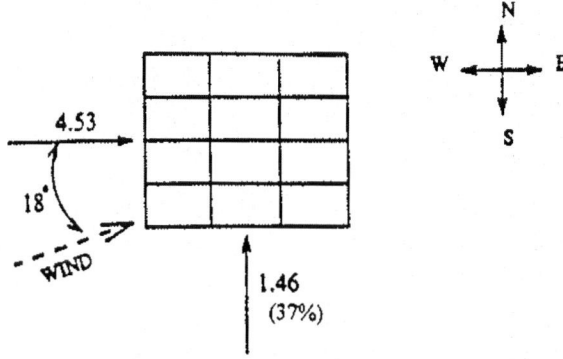

(a) Considering Maximum Load on N-S Frame Set

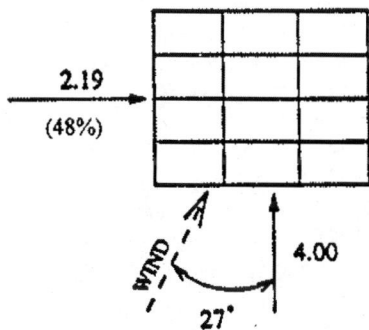

(b) Considering Maximum Load on E-W Frame Set

Figure 5A.5 Force Coefficients, C_f, for Example Design Load Combinations

APPENDIX 5B
HIGH-SOLIDITY OPEN FRAME STRUCTURES

5B.1 Background

The arrangement and congestion of structural members, equipment, piping, electrical items, and appurtenances such as ladders and cages, handrails, stairs, etc. in many open frame process structures creates structures with solidity ratios in excess of those which are the basis for the recommendations in 5.2 and Appendix 5A. See Figure 5B.1 for a typical example of a structure with a dense equipment and piping arrangement. An example of a more open structure for which 5.2 and Appendix 5A would be applicable is shown in Figure 5B.2.

5B.2 Force Coefficients

As the overall solidity of an open frame structure and equipment increases, not only do wind load estimation methods become more cumbersome, but their accuracy may be diminished as the simple, empirical basis for the aerodynamic interaction effects among the framing and equipment becomes insufficient. It is advantageous to work toward a general understanding of the wind loads for high-solidity open frame process structures since accounting for the effects of all of the individual components becomes tedious and less accurate as the structures become more densely occupied. As a result of research performed at Louisiana State University, an analytical model for the force coefficient was derived for porous structures. According to this model, the maximum force coefficient depends on the length to width aspect ratio and the solidity ratio. Equation 5B.1 expresses the maximum force coefficient for a densely occupied open frame structure.

$$C_{max} = \frac{C_0 \cdot \phi \cdot \sqrt{B^2 + L^2}}{2 \cdot B \cdot \left(1 + \frac{B^2}{L^2} - \frac{B}{L^2} \cdot \sqrt{B^2 + L^2}\right)} \quad \text{(Eq. 5B.1)}$$

The variables C_0, ϕ, B, and L in Equation 5B.1 represent an equivalent solid body force coefficient, the total solidity ratio (including equipment), the width of the structure, and the length of the structure, respectively.

Comparisons of the analytical model with wind tunnel measurements for models consisting of multiple, parallel lattice frames and fully three dimensional frameworks showed good agreement in the trends, a low bias, and some scatter. This is illustrated in Figure 5B.3. The experimental data displayed the increase in force coefficient with the plan aspect ratio, which the analytical model predicts. It is possible to describe the upper limit force coefficient only as a function of the length to width ratio. This requires a selection of the empirical parameter, C_0, which ensures that the results from the analytical model adequately envelope the force coefficient for porous structures. A choice of $C_0 = 1.4$ met this criterion for the experimental data examined in the study.

In order to provide a method for estimating an envelope value for the force coefficient for open frame structures, it is proposed that the following relationships be allowed for calculating the gross force coefficients for open frame structures with total projected solidity ratios greater than 50% (including all framing, equipment, vessels, piping, and other appurtenant structures):

$$C_f = \frac{1}{4} \cdot \left(\frac{L}{B}\right)^2 + 1.4 \quad \text{for } L/B < 1.5 \qquad \text{(Eq. 5B.2)}$$

$$C_f = \frac{2}{3} \cdot \left(\frac{L}{B}\right) + 0.9 \quad \text{for } L/B \geq 1.5$$

where L and B are plan dimensions corresponding to the along-wind length and the across-wind width, respectively. The force coefficient is to be applied using the gross area of the structure, A_g.

5B.3 Load Combinations

For this method, the same load cases in 5.2.6 can be used with the exception that F_S should be replaced by F_T, since F_S is not calculated separately in this method. The intent of this method is to facilitate a reasonably conservative design of the structural frame without requiring detailed knowledge of the equipment, which would also allow the structure to be designed to safely accommodate future process changes.

Figure 5B.1 Typical Process Structure with Dense Arrangement of Equipment and Piping

Figure 5B.2 Typical Process Structure with Open Equipment and Piping Arrangement

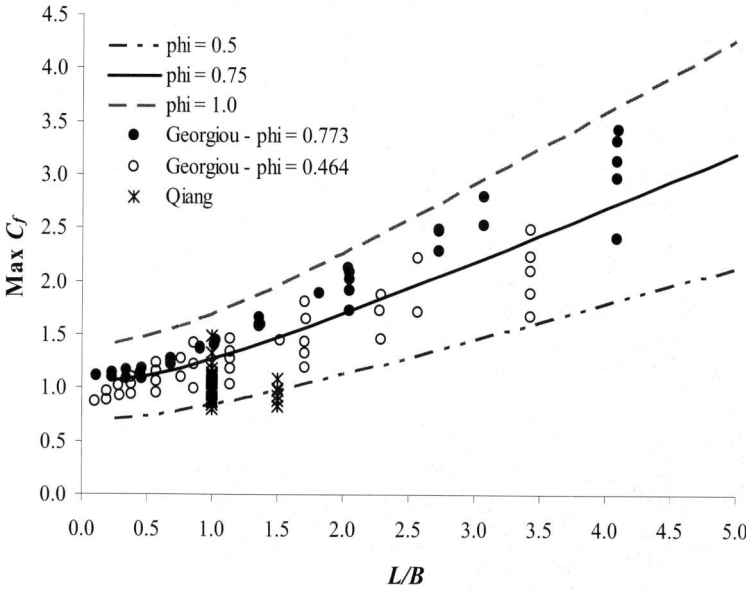

Figure 5B.3 Max. C_f with Respect to Plan Aspect Ratio. The lines indicate the results of Equation 5B.1, and the points represent wind tunnel measurements. (Figure reproduced from Amoroso and Levitan, 2009a)

CHAPTER 6
EXAMPLES

6.0 Introduction

This chapter contains examples of how to apply the recommended guidelines for determining wind loads on the various types of industrial and petrochemical structures.

The site for the example petrochemical facility (may be planned or existing) is located in a rural area along the Texas Gulf Coast about 30 miles inland between Galveston and Corpus Christi. It is surrounded by flat open terrain that extends at least one mile in all directions as shown in Figure 6.1.

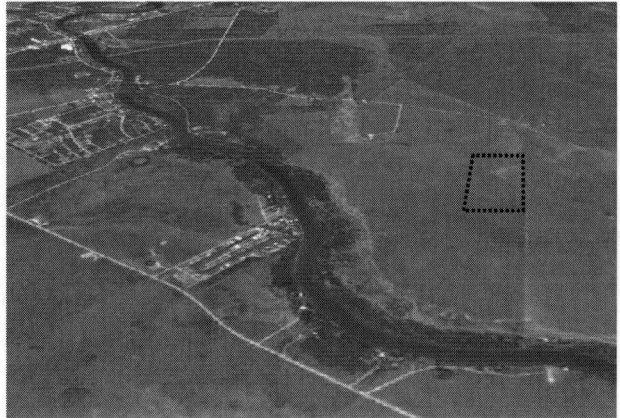

Figure 6.1. Site of Petrochemical Facility (Source: USGS)

The velocity pressure (q_z) is determined by *ASCE 7* equation 6-15. Assume the following design factors per *ASCE 7*:

Basic Wind Speed:	120 mph (3-sec. gust)
Exposure:	C
Exposure Coefficients, K_h and K_z:	Table 6-3 of *ASCE 7*
Topographic Factor, K_{zt}:	1.0
Directionality Factor, K_d:	0.85 for Main Wind Force Resisting System, Components & Cladding, Lattice Frameworks, and Trussed Towers.
Directionality Factor, K_d:	0.95 for Chimneys (Stacks), Tanks, and similar structures (like Vertical Vessels).

WIND LOADS FOR PETROCHEMICAL AND OTHER INDUSTRIAL FACILITIES 111

It should be noted that per *ASCE 7* Section 6.5.4.4, the Directionality Factor is to only be applied when used in conjunction with load combinations specified in Sections 2.3 and 2.4 of *ASCE 7*.

Occupancy Category = II
Importance Factor, I = 1.0

Note: Certain plant structures may be deemed as Occupancy Category III or IV depending on the nature of their operation (e.g., a storage tank containing highly toxic substances). Their Importance Factor will therefore be higher (i.e., 1.15). See Table 1-1 of *ASCE 7* for further clarification.

Gust Effect Factor, $G_f = 0.85$

Note: Certain plant structures (e.g., a partially-filled process tower) may prove to be flexible or dynamically sensitive, thus requiring a rational analysis to determine G_f. See Section 6.5.8 of *ASCE 7* for further clarification.

6.1 Pipe Rack And Pipe Bridge Example

6.1.1 Pipe Rack

The five piping configuration cases considered will be as shown in Figure 6.1.1. Design wind forces are determined by Equation 5.1 (repeated below) where F is the force per unit length of the piping or cable tray:

$$F = q_z G C_f A_e$$

Design wind pressure, for 30 ft. elevation

$q = 0.00256\, K_z K_{zt} K_d V^2 I$ (lb. / ft.^{2})   (*ASCE 7* Section 6.5.10)
$ = 0.613\, K_z K_{zt} K_d V^2 I$ (N / m^2)
$ = 35.4$ psf (1.69 kN / m^2)

Gust effect factor, G = 0.85   (*ASCE 7* Section 6.6.1)

Importance factor I = 1.15 for Category III structures   (*ASCE 7* Table 6-1)

Directionality factor $K_d = 0.85$ (*ASCE 7* Table 6-4)

Force Coefficients

For structural members, $C_f = 1.8$ or alternatively for structural members above the first level $C_f = 1.6$ and below the first level $C_f = 2.0$.   (Section 5.1.3)

112   WIND LOADS FOR PETROCHEMICAL AND OTHER INDUSTRIAL FACILITIES

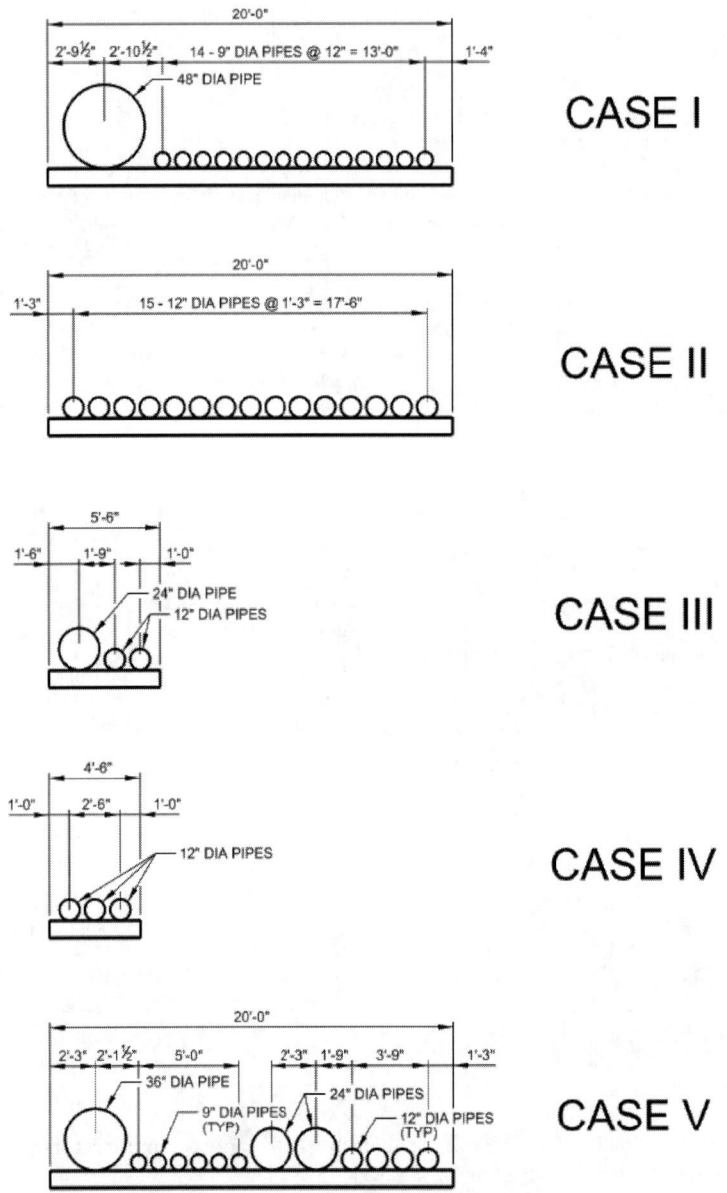

Figure 6.1.1 Pipe Rack Example Cases I - V

WIND LOADS FOR PETROCHEMICAL AND OTHER INDUSTRIAL FACILITIES 113

For pipes, $C_f = 0.7$ (Section 5.1.3)
For cable trays, $C_f = 2.0$

Projected Area

Projected Area per foot of pipe rack, A_e = Largest pipe diameter or cable tray height
+ 10% of the pipe or cable tray width.
(Sections 5.1.1 and 5.1.2)

It is reasonable to limit the projected area of the pipe and cable tray to the maximum tributary height of the structure level.

A_e (total) $\leq A_e$ (max)

Where A_e (total) = A_e (pipes) + A_e (cable tray) and
A_e (max) = the max tributary height of that structure level.

In some cases the pipe rack longitudinal strut or stringer might fall in the shielding envelope and can be deleted from the sum of the projected area of the piping.

PART I - PIPING AND CABLE TRAY

The guidelines require the consideration of the piping or cable tray separately from the structural members. The following calculations are only for piping or cable trays without the structural support members for all five configurations shown in Figure 6.1.1.

Case I - 20 ft. (6.10 m) Wide Rack with one (1) - 48 in. (1.22 m) Pipe and fourteen (14) - 9 in. (0.23 m) Pipes

   Projected Area, A_e = 4 ft. + (20 ft. x 0.10) = 6.0 ft.^{2}/ ft. (1.83 m^2/ m)
   Force per foot, F = 35.4 psf x 0.85 x 0.7 x 6.0 ft.^{2}/ ft. = 126.4 plf (1.84 kN / m)

Case II - 20 ft. (6.10 m) Wide Rack with fifteen (15) - 12 in. (0.30 m) Pipes

   Projected Area, A_e = 1 ft. + (20 ft. x 0.10) = 3 ft.^{2}/ ft. (0.91 m^2/ m)
   Force per foot, F = 35.4 psf x 0.85 x 0.7 x 3.0 ft.^{2}/ ft. = 63.2 plf (0.92 kN / m)

Case III - 5 ft. 6 in. (1.68 m) Wide Rack with one (1) - 24 in. (0.61 m) pipe and two (2) - 12 in. (0.30 m) Pipes

   Projected Area, A_e = 2 ft. + (5.5 ft. x 0.10) = 2.55 ft.^{2}/ ft. (0.77 m^2/ m)
   Force per foot, F = 35.4 psf x 0.85 x 0.7 x 2.55 ft.^{2}/ ft. = 53.7 plf (0.78 kN / m)

Case IV - 4 ft. 6 in. (1.37 m) Wide Rack with three (3) - 12 in. (0.30 m) Pipes

114 WIND LOADS FOR PETROCHEMICAL AND OTHER INDUSTRIAL FACILITIES

Projected Area, A_e = 1 ft. + (4.5 ft. x 0.10) = 1.45 ft.^{2} / ft. (0.44 m^2 / m)
Force per foot, F = 35.4 psf x 0.85 x 0.7 x 1.45 ft.^{2} / ft. = 30.5 plf (0.44 kN / m)

Case V - 20 ft. (6.10 m) Wide Rack with one (1) - 36 in. (0.91 m) Pipe, two (2) - 24 in. (0.61 m) Pipes, four (4) - 12 in. (0.30 m) Pipes, and six (6) - 9 in. (0.23 m) Pipes

Projected Area, A_e = 3 ft. + (20 ft. x 0.10) = 5.0 ft.^{2} / ft. (1.52 m^2 / m)
Force per foot, F = 35.4 psf x 0.85 x 0.7 x 5.0 ft.^{2} / ft. = 105.3 plf (1.54 kN / m)

Cable Trays - 20 ft. (6.10 m) Wide Rack with two (2) - 36 in. (0.91) Trays, one (1) – 24 in. (0.61 m) Tray, two (2) - 18 in. (0.46 m) Trays, two (2) - 12 in. (0.30 m) Trays, and two (2) - 6 in. (0.15 m) Trays, all 6 in. (0.15 m) high.

Projected Area, A_e = 0.5 ft. + (20 ft. x 0.10) = 2.5 ft.^{2} / ft. (0.76 m^2 / m)
Force per foot, F = 35.4 psf x 0.85 x 2.0 x 2.5 ft.^{2} / ft. = 150.5 plf (2.2 kN / m)

PART II - STRUCTURAL MEMBERS

For structural members assume the pipe rack geometry is as follows (see Figure 6.1.2):

- 20 ft. (6.10 m) wide rack with bent spacing on 20 ft. (6.10 m) centers; all columns and stringers not shielded.
- The stringer and the columns are assumed to be W10 and W12 sections respectively. Both stringer and column are assumed to have a level of fireproofing insulation as indicated in the projected area.
- Stringer member projected area per foot of rack = 1 ft.^{2} / ft. (0.31 m^2 / m)
- Column projected area per foot of column = 1.25 ft.^{2} / ft. (0.38 m^2 / m).
- Three levels of pipes and cable trays at elevations 18 ft. (5.49 m), 24 ft. (7.32 m), and 30 ft. (9.15 m).
- One level of struts at 21 ft. (6.40 m).

q_{18} [at elevation 18 ft. (5.49 m)] = 31.8 psf (1.52 kN / m^2)
q_{21} [at elevation 21 ft. (6.40 m)] = 32.8 psf (1.57 kN / m^2)
q_{24} [at elevation 24 ft. (7.32 m)] = 33.8 psf (1.62 kN / m^2)
q_{30} [at elevation 30 ft. (9.15 m)] = 35.4 psf (1.69 kN / m^2)

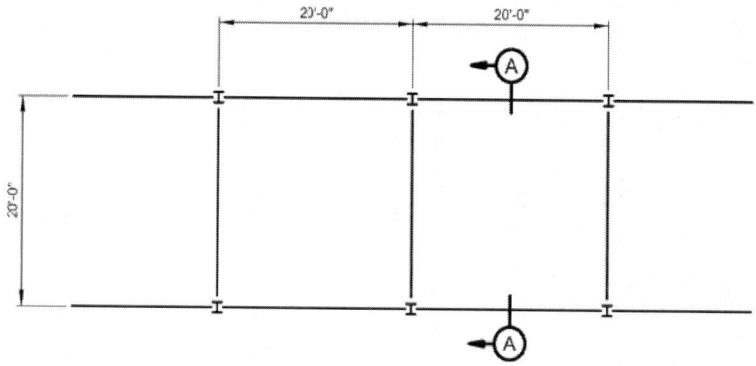

PIPE RACK PLAN

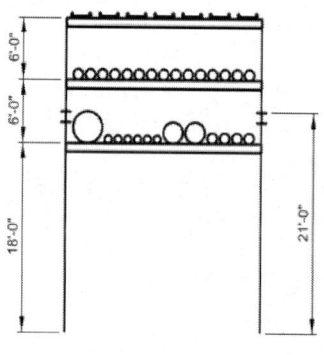

SECTION

Figure 6.1.2 Pipe Rack Example Framing Figure

Next calculate wind loads per bent using a force coefficient $C_f = 1.8$ for all members per Section 5.1.3. Projected area for stringers at elevation 21 ft. (6.40 m) is calculated as the sum of stringers times the stringer depth times the bent spacing.

$$\text{Projected Area of Stringers} = 2 \text{ stringers} \times 1 \text{ ft. depth} \times 20 \text{ ft. bent spacing}$$
$$= 40 \text{ ft.}^2 \ (3.72 \text{ m}^2)$$

Similarly the projected area for columns is:

$$\text{Projected Area of Columns} = 2 \text{ columns} \times 1.25 \text{ ft. width} \times 30 \text{ ft. height}$$
$$= 75 \text{ ft.}^2 \ (6.97 \text{ m}^2)$$

Total force on the structural members is per Equation 5.1:

$$F = 32.8 \text{ psf} \times 0.85 \times 1.8 \times 40 \text{ ft.}^2 + 35.4 \text{ psf} \times 0.85 \times 1.8 \times 75 \text{ ft.}^2$$
$$= 6070 \text{ lbs. } (27.0 \text{ kN})$$

Alternatively, using $C_f = 1.6$ for members above the first level and $C_f = 2.0$ for members below the first level:

Projected area for stringers $= 40 \text{ ft.}^2 \ (3.72 \text{ m}^2)$

Projected area for columns above first level $= 2$ columns x 1.25 ft. x 12 ft. high
$$= 30 \text{ ft.}^2 \ (2.79 \text{ m}^2)$$

Projected area for columns below first level $= 2$ columns x 1.25 ft. x 18 ft. high
$$= 45 \text{ ft.}^2 \ (4.18 \text{ m}^2)$$

Total force on structure $F = (32.8 \text{ psf} \times 40 \text{ ft.}^2 + 35.4 \text{ psf} \times 30 \text{ ft.}^2) \times 0.85 \times 1.6 +$
$$31.8 \text{ psf} \times 0.85 \times 45 \text{ ft.}^2 \times 2.0$$
$$= 5661 \text{ lbs. } (25.2 \text{ kN})$$

SUMMARY AND CONCLUSION

To combine the effects of the piping, cable tray and the structural members, the pipe rack structure along with Case V pipe configuration on the bottom level, Case II pipe configuration on the middle level, and cable trays on the top level are used. See Figure 6.1.2.

Calculate the Total Base Shear for the structure.

Recalculate the wind load from Part 1 for the appropriate elevation shown in Figure 6.1.2.

Bottom Level of Pipes 105.3 plf x (31.8 / 35.4) x 20 ft. = 1892 lbs. (8.4 kN)

WIND LOADS FOR PETROCHEMICAL AND OTHER INDUSTRIAL FACILITIES   117

Middle Level of Pipes 63.2 plf x (33.8 / 35.4) x 20 ft. = 1207 lbs. (5.4 kN)

Top Level of Cable Trays 150.5 plf x 20 ft. = 3010 lbs. (13.4 kN)

Structure = 6070 lbs. (27.0 kN)

Total Base Shear per bent = 12179 lbs. (54.2 kN)

6.1.2 Pipe Bridge Example

The pipe bridge case is similar to the pipe rack as described in Section 6.1.1. Design wind forces are determined by Equation 5.1 and 5.1a (repeated below) where F is the force per unit length of the piping or cable tray:

$F = q_z GC_f A_e$

Gust effect factor, $G = 0.85$	(*ASCE 7* Section 6.5.8)
Importance factor $I = 1.15$ for Category III structures	(*ASCE 7* Table 6-1)
Directionality factor $K_d = 0.85$	(*ASCE 7* Table 6-4)

BRIDGE STRUCTURAL MEMBERS

For structural members assume the pipe bridge geometry is as follows (see Figure 6.1.3):

- 20 ft (6.10 m) wide x 80 ft (24.4 m) long bridge with pipe support bents at 20 ft (6.10 m) centers.
- The top chord, bottom chord and pipe support beams are assumed to be W10, the web columns are assumed to be W8, and the diagonals are assumed to be WT4. The pipe bridge is not fireproofed.
- Chord member projected area per foot of bridge = 0.84 ft^2/ft (0.26 m^2/m)
- Column projected area per foot of bridge = 0.67 ft^2/ft x 12 ft/10 ft = 0.80 ft^2/ft (0.24 m^2/m)
- Diagonal member projected area per foot of bridge = 0.34 ft^2/ft x 15.6 ft/10 ft = 0.53 ft^2/ft (0.16 m^2/m)
- Three levels of pipes and cable trays at elevations 24 ft (7.32 m), 30 ft (9.15 m) and 36 ft (11.0 m).

q_{24} [at elevation 24 ft. (7.32 m)] = 33.8 psf (1.62 kN/m^2)
q_{30} [at elevation 30 ft. (9.15 m)] = 35.4 psf (1.70 kN/m^2)
q_{36} [at elevation 36 ft. (11.0 m)] = 36.8 psf (1.76 kN/m^2)

118 WIND LOADS FOR PETROCHEMICAL AND OTHER INDUSTRIAL FACILITIES

Next calculate wind loads on bridge members using a force coefficient $C_f = 1.6$ for all members per Section 5.1.3. In this example only members in the truss area are considered, and wind loads on column projected area below the bridge would need to be calculated for total wind loading.

A. LATERAL LOAD ON PIPE BRIDGE

Lateral Load on Bridge Members

All steel members are considered not to be shielded.

Projected area of half height
of bridge, $A_s = 2 \times (0.84 \text{ ft/ft} + 0.8 \text{ ft}^2/\text{ft}/2 + 0.53 \text{ ft}^2/\text{ft}/2)$
$= 3.01 \text{ ft}^2/\text{ft} \ (0.91 \text{ m}^2/\text{m})$

Force on top chord, $F = 36.8 \text{ psf} \times 0.85 \times 1.6 \times 3.0 \text{ ft}^2/\text{ft} \times 80 \text{ ft}$.
$= 12,052 \text{ lbs. } (53.6 \text{ kN})$

Force on bottom chord, $F = 33.8 \text{ psf} \times 0.85 \times 1.6 \times 3.0 \text{ ft}^2/\text{ft} \times 80 \text{ ft}$
$= 11,069 \text{ lbs. } (49.2 \text{ kN})$

The lateral loads on bridge members, $F = 12,052 \text{ lbs.} + 11,069 \text{ lbs.}$
$= 23,121 \text{ lbs. } (102.9 \text{ kN})$

Alternatively, apply Section 5.2 for Open Frame Structures to generate lateral loads on the bridge. In this case shielding is considered.

For bridge lateral surface:
$A_g = 12 \text{ ft}^2/\text{ft} + 0.84 \text{ ft}^2/\text{ft} = 12.84 \text{ ft}^2/\text{ft} \ (3.91 \text{ m}^2/\text{m})$

The effective solid area: $A_s = 0.84 \text{ ft}^2/\text{ft} \times 2 + 0.80 \text{ ft}^2/\text{ft} + 0.53 \text{ ft}^2/\text{ft}$
$= 3.01 \text{ ft}^2/\text{ft} \ (0.92 \text{ m}^2/\text{m})$
$= A_e$

The solidity ratio: $\varepsilon = A_s/A_g$
$= 3.01 \text{ ft}^2/\text{ft}/12.84 \text{ ft}^2/\text{ft} = 0.234$
$N = 2$
$S_f = 20 \text{ ft. } (6.1 \text{ m})$
$B = 80 \text{ ft. } (24.4 \text{ m})$
$S_f/B = 20/80 = 0.25$

From Figure 5.1, $C_{Dg} = 0.60$

From Equation 5.2, $C_f = C_{Dg}/E = 0.60/0.234 = 2.56$

The lateral loads on bridge members,

WIND LOADS FOR PETROCHEMICAL AND OTHER INDUSTRIAL FACILITIES

$$F = q_z \, G \, C_f \, A_e \quad \text{(Eq. 5.1a)}$$
$$= (36.8 \text{ psf} + 33.8 \text{ psf})/2 \times 0.85 \times 2.56 \times 3.01 \text{ ft}^2/\text{ft} \times 80 \text{ ft}$$
$$= 18{,}496 \text{ lbs } (82.3 \text{ kN})$$

Lateral Load on Pipes and Cable Trays

The pipe support beams in the bridge are similar to those in the pipe rack example. Case V pipes on the bottom level, Case II pipes on the middle level, and cable trays on the top level are used. See Figure 6.1.3.

Bottom Level of Pipes $F = 105.3$ plf x $(33.8/35.4)$ x 80 ft. = 8043 lbs (35.8 kN)

Middle Level of Pipes $F = 63.2$ plf x $(35.4/35.4)$ x 80 ft. = 5056 lbs (22.5 kN)

Top Level of Cable Trays $F = 150.5$ plf $(36.8/35.4)$ x 80 ft. = 12516 lbs (55.7 kN)

Total Lateral Load on Pipe Bridge

Using the first of the two methods for calculating the force on the structure,

$$F = 23{,}121 \text{ lbs} + 8043 \text{ lbs} + 5056 \text{ lbs} + 12{,}516 \text{ lbs}$$
$$= 48{,}736 \text{ lbs } (216.8 \text{ kN})$$

Alternatively, the second method for calculating the force on the structure could be used.

B. LONGITUDINAL LOAD ON PIPE BRIDGE

Longitudinal Load on Bridge Members

Apply Section 5.2 for Open Frame Structures to generate longitudinal loads for bridge members using an interior truss bay to define the projected area.

For bridge cross section: $A_g = 20.83$ ft x 12.83 ft = 267.3 ft^2 (24.8 m^2)

The solid area of bridge frame: $A_{s1} = 0.83$ ft x 20 ft x 3 + 0.67 ft x 12 ft x 2
$$= 65.9 \text{ ft}^2 \, (6.1 \text{ m}^2)$$

The solid projected area of diagonal bracing: $A_{s2} = 0.33$ ft x 12 ft x 2
$$= 7.92 \text{ ft}^2 \, (0.74 \text{ m}^2)$$

The effective solid area: $A_s = A_{s1} + A_{s2}$
$$= 65.9 \text{ ft}^2 + 7.92 \text{ ft}^2 = 73.8 \text{ ft}^2 \, (6.86 \text{ m}^2)$$
$$= A_e$$

The solidity ratio: $\varepsilon = A_s/A_g$
$= 73.8 \text{ ft}^2/267.3 \text{ ft}^2 = 0.276$
$N = 9$
$S_f = 10 \text{ ft} (3.1 \text{ m})$
$B = 20.83 \text{ ft} (6.4 \text{ m})$
$S_f/B = 10/20.83 = 0.48$

From Figure 5.1, $C_{Dg} = 2.18$

From Equation 5.2, $C_f = C_{Dg}/\varepsilon = 2.18/0.276 = 7.89$

The longitudinal loads on bridge members,

$$F = q_z G C_f A_e \quad \text{(Eq. 5.1a)}$$
$= (36.8 \text{ psf} + 33.8 \text{ psf})/2 \times 0.85 \times 7.89 \times 73.8 \text{ ft}^2$
$= 17,471 \text{ lbs} (77.7 \text{ kN})$

The shielding of pipes at the end of the pipe bridge will reduce the longitudinal wind load on the bridge structure. But at present, proven laboratory testing to determine the shielding results is lacking. Therefore, this shielding effect has not been addressed in this example.

Longitudinal Load on Pipes and Cable Trays

Assume all piping is turned up from the pipe rack to the pipe bridge at the beginning of the bridge and returns down to the pipe rack at the end of the bridge. Also assume half of the pipe and cable tray loads are applied to the pipe bridge and the other half to the pipe rack.

Bottom Level of Pipes:

The solid area of pipes = (36 in/12 in/ft + 2 × 24 in/12 in/ft + 4 × 12 in/12in/ft
+ 6 × 9 in/12 in/ft) × 6 ft = 93.0 ft^2 (8.6 m^2)

The center-to-center distance of pipes is less than 3 diameters, therefore C_f should be increased 20% per 5.4.1.4,

$F = 33.8 \text{ psf} \times 0.85 \times 0.7 \times 1.2 \times 93.0 \text{ ft}^2 \times 2 = 4489 \text{ lbs} (20.0 \text{ kN})$

Middle Level of Pipes:

The solid area of pipes = (15 × 12 in/12 in/ft) × 6 ft = 90.0 ft^2 (8.4 m^2)

$F = 35.4 \text{ psf} \times 0.85 \times 0.7 \times 1.2 \times 90.0 \text{ ft}^2 \times 2 = 4550 \text{ lbs} (20.2 \text{ kN})$

Top Level of Cable Trays: assume cable trays and cables occupied 90% of the cross-sectional area of the pipe rack and $C_f = 1.0$ in this case.

The solid area of cable trays = 0.9 x 20 ft x 6 ft = 108.0 ft² (10.0 m²)

F = 36.8 psf x 0.85 x 1.0 x 108 ft² x 2 ft = 6757 lbs (30.1 kN)

Total Longitudinal Load on Pipe Bridge

F = 17,471 lbs + 4489 lbs/2 + 4550 lbs/2 + 6757 lbs/2
= 25,369 lbs (112.9 kN)

122 WIND LOADS FOR PETROCHEMICAL AND OTHER INDUSTRIAL FACILITIES

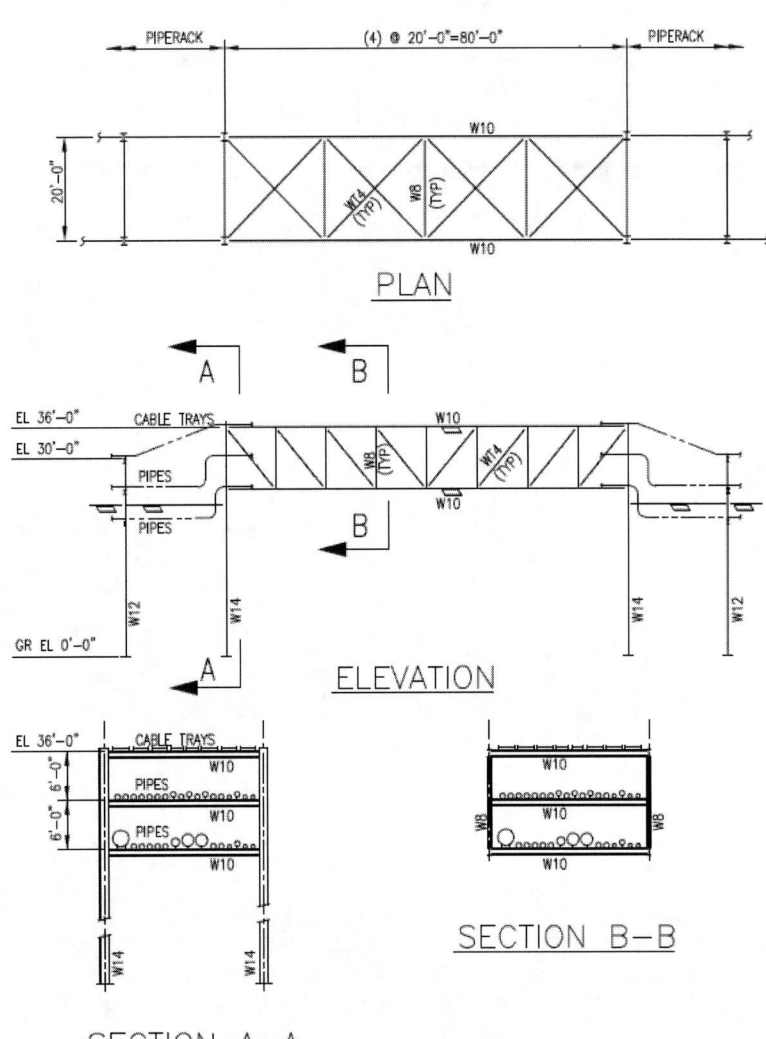

Figure 6.1.3 Pipe Bridge Example

WIND LOADS FOR PETROCHEMICAL AND OTHER INDUSTRIAL FACILITIES

6.2 Open Frame Examples

In this section, the wind load will be determined on an open frame structure using the three methods described in Section 5.2, Appendix 5A, and Appendix 5B.

The structure considered is shown in Figures 6.2.1, 6.2.2, and 6.2.3. Design wind forces are determined by Equation 5.1a:

$$F = q_z G C_f A_e$$

The velocity pressure q_z is determined using Section 6.5 of *ASCE 7*.

$$q_z = 0.00256 K_z K_{zt} K_d V^2 I \qquad (ASCE\ 7\ Eq.\ 6\text{-}15)$$

Refer to Chapter 6 (Introduction) for the values to be used for the example. (The structure will be assumed to be Occupancy Category II with an importance factor I = 1.0). It is convenient to determine the velocity pressures at the mid-floor heights and at the top of the structure. K_z and q_z are determined and summarized in Table 6.2.1.

Table 6.2.1 q_z and K_z

Height above Ground z(ft.)	K_z	q_z (psf)
10	0.85	30.6
34	1.00	36.0
65	1.15	41.4
h = 83	1.22	44.0

Note: To convert psf to N/m^2, multiply values in this table by 47.878.

Although the top of the third floor level is at 82 ft., the structure height h was increased slightly to account for the handrail and minor equipment on top of the structure (see Figures 6.2.2 and 6.2.3).

The gust effect factor is determined next. The ratio of height/least horizontal dimension = 83 ft. / 41 ft. = 2.02 < 4, therefore the structure is not considered a flexible structure. Use G = 0.85, as described at the beginning of Chapter 6.

124   WIND LOADS FOR PETROCHEMICAL AND OTHER INDUSTRIAL FACILITIES

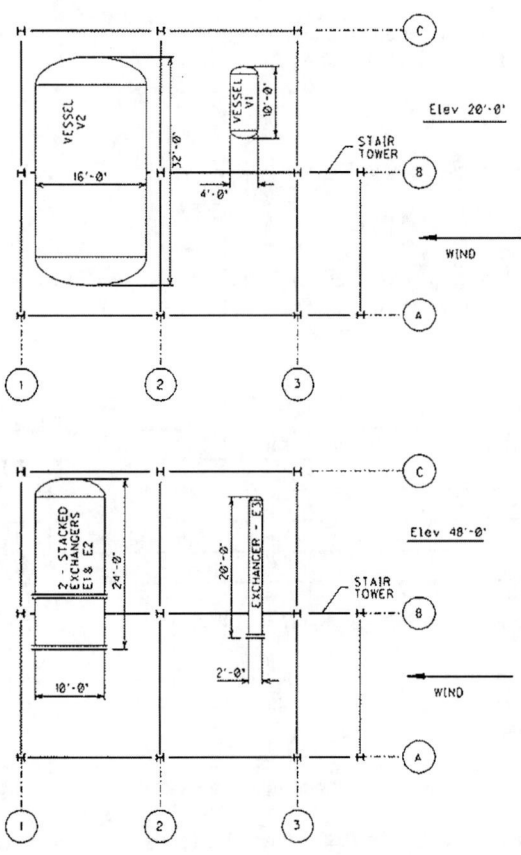

Figure 6.2.1 Example for Open Frame Structure – Arrangement Plan

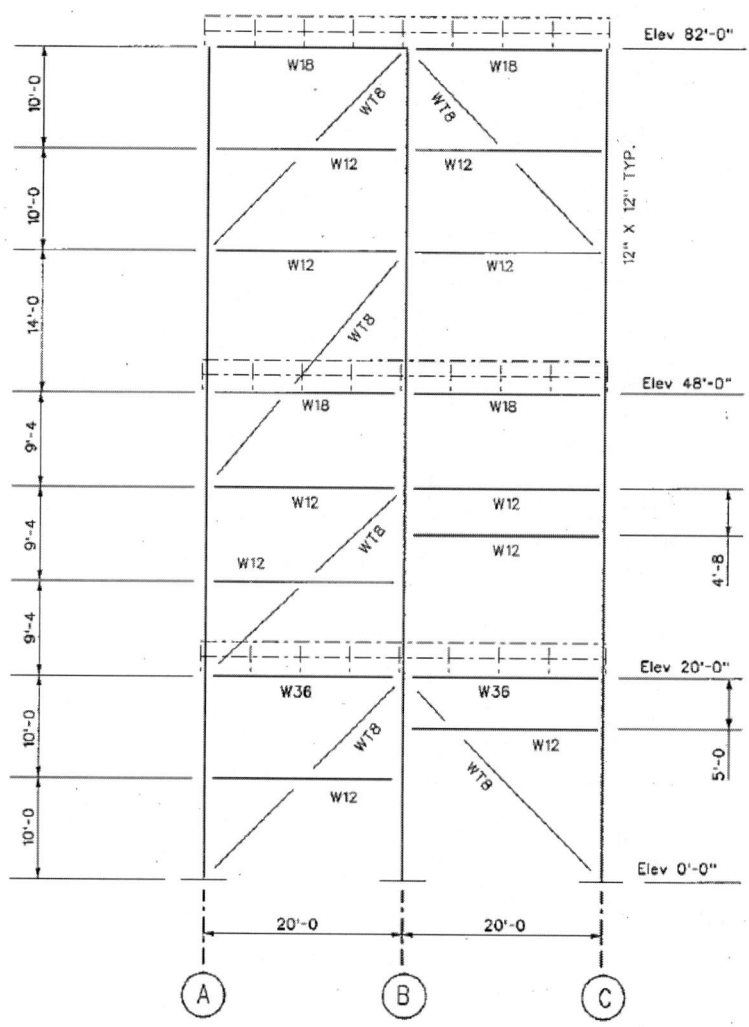

Figure 6.2.2 Example for Open Frame Structure – Column Line 3

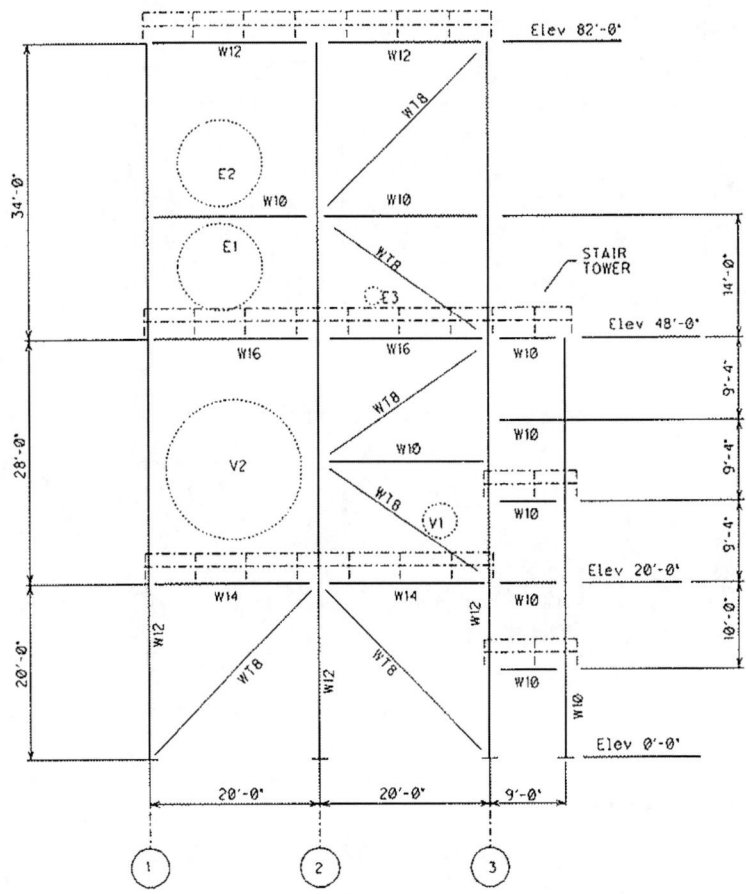

Figure 6.2.3 Example for Open Frame Structure – Column Line A

WIND LOADS FOR PETROCHEMICAL AND OTHER INDUSTRIAL FACILITIES

6.2.1 Open Frame Example – Using Method of Section 5.2

6.2.1.1 Along Wind Force Calculations (Wind Toward Frame 3)

In order to calculate the force coefficient, the solidity ratio ε must first be computed from Equation 5.3, which stated $\varepsilon = A_S / A_g$. The gross area (or envelope area) is the area within the outmost projections of the front face normal to the wind direction. Note that the width used below is measured from outside column face to outside column face. For the wind direction shown in Figure 6.2.1,

$$A_g = 83 \text{ ft. (height)} \times 41 \text{ ft. (width)} = 3{,}403 \text{ ft.}^2 \; (316 \text{ m}^2)$$

To determine the effective solid area, the solid area of the windward frame must first be calculated per 5.2.4.1. In order to facilitate the computation of forces later in the problem, it is convenient to calculate the solid areas from mid-floor to mid-floor, and sum these to obtain the total solid area of the frame. Member sizes are given in the respective figures (except details of handrails and stairs). Calculation of the solid area of the windward frame (column line 3) is summarized in Table 6.2.1.1. The stairs were considered as part of the windward frame (see Figure 6.2.1). The stairs column in Table 6.2.1.1 includes areas of stair stringers, struts, handrails, and bracing.

Table 6.2.1.1 Solid Area of Windward Frame and Vertical Bracing in Bents Parallel to Wind Direction (Wind toward Frame 3)

Floor Level	Tributary Height (ft.)	Solid Areas (ft.^{2})							
		Cols.	Beams	Intermed. Beams	Bracing (Normal)	Bracing (3 Parallel Bents)	Handrails	Stairs	Total
0	0 - 10 = 10	30	0	20	19	20	0	76	165
1	10 - 34 = 24	72	120	60	33	48	32	150	515
2	34 - 65 = 31	93	60	80	31	62	32	91	449
3	65 - 83 = 18	51	60	40	32	34	32	0	249
							Total Solid Area of Windward Frame and Parallel Bents (ft.^{2}) =		1378 (128 m^2)

Note: To convert ft.^{2} to m^2, multiply values in this table by 0.0929.

Since the middle and leeward frames (column lines 2 and 1, respectively) are similar to the windward frame with the exception of not having stairs, the solid areas and hence the solidity ratios for these two frames will be less than the windward frame, so A_S is equal to the solid area of windward frame per 5.2.4.4, which leads to

$$\varepsilon = A_S / A_g = 1{,}378 \text{ ft}^2 / 3{,}403 \text{ ft}^2 = 0.405$$

Next, the coefficient C_{Dg} is obtained from curves given in Figure 5.1 as a function of the solidity ratio ε, the number of frames N, and the frame spacing ratio S_F/B. As defined in Figure 5.1, N=3 and S_F/B = 20 ft / 41 ft = 0.488. From Figure 5.1, for N=3 and extrapolating as required for $\varepsilon = 0.405$

$C_{Dg} = 1.18$, for $S_F / B = 0.5$
$C_{Dg} = 1.12$, for $S_F / B = 0.33$

Interpolating for $S_F / B = 0.488$

$$C_{Dg} = 1.12 + (1.18 - 1.12)[(0.488 - 0.33)/(0.5 - 0.33)] = 1.176$$

Next, the gross area force coefficient C_{Dg} is converted into a force coefficient compatible with *ASCE 7* by means of Equation 5.2.

$$C_f = C_{Dg} / \varepsilon = 1.176 / 0.405 = 2.90$$

The force coefficient could also have been determined directly using the alternate method of Appendix 5A. The alternate method is somewhat more time consuming, but it covers a wider range of solidity ratios and frame spacing ratios, and can also be used to determine more accurate load combinations. (See Section 6.2.2)

The area of application of force A_e has already been determined per floor level during calculation of solidity ratio. The wind force transmitted to each floor level may now be found by Equation 5-1a, $F = q_z G C_f A_e$, as shown below. For floor levels which have solid flooring, the wind load on these levels may be reduced by the shielding factor η_{floor}, per 5.2.4.2. For this example problem, assume that the lower two levels are decked with checkered plate (solid flooring) and that the top level is floored with grating. Applying Equation 5.4 -- $\eta_{floor} = 1 - 0.2 (A_{fb} / A_S)$:

Level 1 -- $A_{fb} = 120$ ft^2 $A_S = 515$ ft^2 $\eta_{floor} = 0.953$
Level 2 -- $A_{fb} = 60$ ft^2 $A_S = 449$ ft^2 $\eta_{floor} = 0.973$
Level 3 -- $\eta_{floor} = 1.0$

The total force on the structural frame and appurtenances, F_S, is 127.6 kips (567.6 kN), found by summing the forces at all levels in Table 6.2.1.2.

Table 6.2.1.2 Total Force - Structural Frame and Appurtenances - F_S (Wind toward Frame 3)

Floor Level	q_z (psf)	G	C_f	A_e (ft^2)	η_{floor}	F (lbs)
0	30.6	0.85	2.90	165	--	12,446
1	36.0	0.85	2.90	515	0.953	43,553
2	41.4	0.85	2.90	449	0.973	44,584
3	44.0	0.85	2.90	249	1.0	27,007
					$F_S = \sum F =$	127,590

Note: To convert pounds force (lbs) to newtons (N), multiply F values in this table by 4.448.

These forces are due to wind acting on the frame only. Wind forces acting on the vessels, equipment and piping are computed in accordance with Sections 5.1 and 5.4 and added at the levels where the items are located. The structure supports two horizontal vessels on floor level one and three horizontal exchangers on floor level two (see respective figures). All of these items are subject to transverse wind loads for the wind direction under consideration. The projected diameter is equal to the vessel insulated diameter plus 1.5 ft. (0.46 m) per 5.4.2.2. The force coefficient is determined from *ASCE 7* Figure 6-21. The vessels and exchangers were assumed to be moderately smooth. These properties are listed in Table 6.2.1.3.

Table 6.2.1.3 Equipment Properties (Wind toward Frame 3)

Equip.	Floor Level	Vessel Dia. (ft)	Proj. Dia. D (ft)	Length B (ft)	Proj. Area A_e (ft^2)	B/D	C_f
V1	1	4	5.5	10	55	1.8	0.51
V2	1	16	17.5	32	560	1.8	0.51
E1	2	10	11.5	24	276	2.1	0.52
E2	2	10	11.5	24	276	2.1	0.52
E3	2	2	3.5	20	70	5.7	0.58

Note: To convert ft.^{2} to m^2, multiply A_e values in this table by 0.0929.

The wind load on a given vessel or exchanger is the sum of the wind load on the supports, platforms, large connecting pipes, and the cylinder itself as summarized in Table 6.2.1.4. This example problem only considers the load on the cylinder. For determining q_h, the height of each vessel was assumed to be the mid-floor elevation above the supporting floor level. A small improvement in accuracy could be obtained by using the actual top elevation for each piece of equipment.

Table 6.2.1.4 Gross Wind Forces – Equipment (Wind toward Frame 3)

Equipment	q_h (psf)	G	C_f	A_e (ft^2)	F (lbs)
V1	36.0	0.85	0.51	55	858
V2	36.0	0.85	0.51	560	8739
E1	41.4	0.85	0.52	276	5050
E2	41.4	0.85	0.52	276	5050
E3	41.4	0.85	0.58	70	1429

Note: To convert pounds force (lbs) to newtons (N), multiply F values in this table by 4.448.

The projected area of piping and electrical was given as 20% of the projected vessel area. For example, on floor level one, the piping area is equal to 0.2 (55 + 560) = 123 ft^2 (11.42 m^2). The wind load on piping per floor level is summarized in Table 6.2.1.5.

Table 6.2.1.5 Gross Wind Forces - Piping per Level (Wind toward Frame 3)

Floor Level	q_h (psf)	G	C_f	A_e (ft^2)	F (lbs)
1	36.0	0.85	0.7	123	2635
2	41.4	0.85	0.7	124	3054

Note: To convert pounds force (lbs) to newtons (N), multiply F values in this table by 4.448.

The total equipment load per floor level is equal to the sum of all of the vessel, exchanger, and piping wind loads on that floor level. For purposes of determining the overall wind load on the structure, equipment and piping loads can be reduced to account for shielding effects (shielding of the equipment by upwind portions of the structure, shielding of portions of the structure by upwind equipment, and equipment to equipment shielding). Note that for purposes of designing individual vessels and supports, the loads should not be reduced.

Since the vessels and exchangers are in the wind shadow of the stairs, vertical bracing, middle column and intermediate beams, it is appropriate to reduce the equipment load for shielding by the upwind frames. Per Equation 5.5, the shielding factor η_{equip} is given by

$$\eta_{equip} = \exp[-1.4\,(C_f \cdot \varepsilon)^{1.5}]$$

where $\varepsilon = 0.405$ and $C_f = 2.90$, as calculated previously

$$\eta_{equip} = \exp[-1.4\,(2.90 \times 0.405)^{1.5}] = 0.17$$

Note that if the solidity of the upwind frames varies considerably from level to level, it would be appropriate to calculate an ε and C_f for each level supporting equipment rather than using a single value of ε and C_f for the overall structure.

Summing the loads on the vessels, exchangers, and piping per level and applying the shielding factor yields the total wind load on equipment shown in Table 6.2.1.6.

Table 6.2.1.6 Gross Wind Force - Equipment and Piping (Wind toward Frame 3)

Floor Level	Equipment and Piping Load	η_{equip}	Reduced Load - lbs (kN)
1	V1 + V2 + piping = 858 + 8739 + 2635 = 12,232 lbs. (54.4 kN)	0.17	2079 (9.2)
2	E1 + E2 + E3 + piping = 5050 + 5050 + 1429 + 3054 = 14,583 (64.9 kN)	0.17	2479 (11.0)
	Total Equipment Load = $\sum$ Reduced Loads =		4558 (20.2)

WIND LOADS FOR PETROCHEMICAL AND OTHER INDUSTRIAL FACILITIES   131

6.2.1.2 Crosswind Force Calculations (Wind Toward Frame A)

The next step is to repeat the analysis for the nominal wind direction normal to column line A (see Figures 6.2.1 and 6.2.3). The member sizes are the same on this elevation except that the intermediate beams are W10's and

Beams EL 20 ft - 0 in -- W14
Beams EL 48 ft - 0 in -- W16
Beams EL 82 ft - 0 in -- W12

The gross area of the windward face includes the stair tower on the right hand side of the structure.

$$A_g = (83 \times 41) + (9 \times 49) = 3,844 \text{ ft}^2$$

The solid areas for the windward frame are given below. The stairs column in the table includes the areas of the stair column, struts, and handrails. (See Table 6.2.1.7.)

Table 6.2.1.7 Solid Area of Windward Frame and Vertical Bracing in Bents Parallel to Wind Direction (Wind toward Frame A)

Floor Level	Tributary Height (ft.)	Solid Areas (ft.^{2})							
		Cols.	Beams	Intermed. Beams	Bracing (Normal)	Bracing (3 Parallel Bents)	Handrails	Stairs	Total
0	0 - 10 = 10	30	0	0	19	20	0	24	93
1	10 - 34 = 24	72	46	8	35	48	32	44	285
2	34 - 65 = 31	93	53	41	36	62	32	36	353
3	65 - 83 = 18	51	40	0	16	34	32	0	173
						Total Solid Area of Windward Frame and Parallel Bents (ft.^{2}) =			904 (84 m^2)

Note: To convert ft^2 to m^2, multiply values in this table by 0.0929.

Since the solidity of neither the middle nor leeward frames (column lines B and C, respectively) exceeds that of the windward frame, A_S is equal to the solid area of windward frame, yielding

$$\varepsilon = A_S / A_g = 904 \text{ ft}^2 / 3,844 \text{ ft}^2 = 0.235$$

The frame spacing ratio in this direction is $S_F / B = 20 \text{ ft} / 46 \text{ ft} = 0.435$. Since the width is not uniform (the stair tower stops at the second floor level), an average value of B was used. From Figure 5.1 for $N = 3$ and $\varepsilon = 0.235$

$C_{Dg} = 0.82$ for $S_F / B = 0.5$
$C_{Dg} = 0.79$ for $S_F / B = 0.33$

Therefore use $C_{Dg} = 0.809$

$C_f = C_{Dg} / \varepsilon = 0.809 / 0.235 = 3.44$

As was done for the orthogonal direction, the shielding factor, η_{floor}, is determined for each floor level by applying Equation 5.4 -- $\eta_{floor} = 1 - 0.2\,(A_{fb} / A_s)$.

Level 1 -- $A_{fb} = 46\ ft^2$ $A_S = 285\ ft^2$ $\eta_{floor} = 0.968$
Level 2 -- $A_{fb} = 53\ ft^2$ $A_S = 353\ ft^2$ $\eta_{floor} = 0.970$
Level 3 -- $\eta_{floor} = 1.0$

The wind forces per floor level are shown in Table 6.2.1.8.

The wind direction is parallel to the axis of the vessels and exchangers (longitudinal wind). The vessels have rounded heads and the exchangers have flat heads. Force coefficients for this case are given in 5.4.2.4, and the wind loads are tabulated in Table 6.2.1.9.

Table 6.2.1.8 Total Force - Structural Frame and Appurtenances - F_S (Wind toward Frame A)

Floor Level	q_z (psf)	G	C_f	A_e (ft^2)	η_{floor}	F (lbs)
0	30.6	0.85	3.44	93	--	8,321
1	36.0	0.85	3.44	285	0.968	29,040
2	41.4	0.85	3.44	353	0.970	41,450
3	44.0	0.85	3.44	173	1.0	22,257
					$F_S = \sum F =$	101,068

Note: To convert pounds force (lbs) to newtons (N), multiply F values in this table by 4.448.

Table 6.2.1.9 Gross Wind Forces – Equipment (Wind Toward Frame A)

Equipment	q_h (psf)	G	C_f	A_e (ft^2)	F (lbs)
V1	36.0	0.85	0.5	24	367
V2	36.0	0.85	0.5	241	3687
E1	41.4	0.85	1.2	104	4392
E2	41.4	0.85	1.2	104	4392
E3	41.4	0.85	1.2	10	422

Note: To convert pounds force (lbs) to newtons (N), multiply F values in this table by 4.448.

The wind load on the piping and electrical is the same as calculated previously. For this wind direction, there is no significant shielding of the equipment by the windward frame and no equipment to equipment shielding, so no reduction is taken on equipment load. Summing the loads on the vessels, exchangers, and piping per level yields

Level 1: V1 + V2 + piping = 367 + 3687 + 2635 = 6689 lbs (29.8 kN)
Level 2: E1 + E2 + E3 + piping = 4392 + 4392 + 422 + 3054 = 12,260 lbs (54.4 kN)

WIND LOADS FOR PETROCHEMICAL AND OTHER INDUSTRIAL FACILITIES 133

for a total equipment and piping load of F_E = 18,949 lbs. (84.3 kN).

6.2.1.3 Open Frame Example (Using Method of Section 5.2) – Summary and Conclusion

The results thus far are summarized in Table 6.2.1.10. The load combinations for design are the application of F_T in one direction simultaneously with 0.5 F_S in the other, per 5.2.6.1. These combinations are shown in Figure 6.2.1.1.

Table 6.2.1.10 Summary

	Wind -- Direction 1	Wind -- Direction 2
Wind Load on Structural Frame F_S	127.6 kips (568 kN)	101.1 kips (450 kN)
Wind Load on Equipment and Piping F_E	4.6 kips (20 kN)	18.9 kips (84kN)
Total Wind Load on Structure F_T	132.2 kips (588 kN)	120.0 (534 kN)

It should be noted that this example problem only determines the <u>total</u> wind load in each direction. For actual design of the structure, the wind load calculation must be refined. The wind load on the structure frame (F_S) in each direction would need to be determined for each structure node by considering A_e for each individual node. The wind load on equipment and piping (F_E) would need to be applied as appropriate to each individual item of equipment. This refinement is beyond the scope of this example problem.

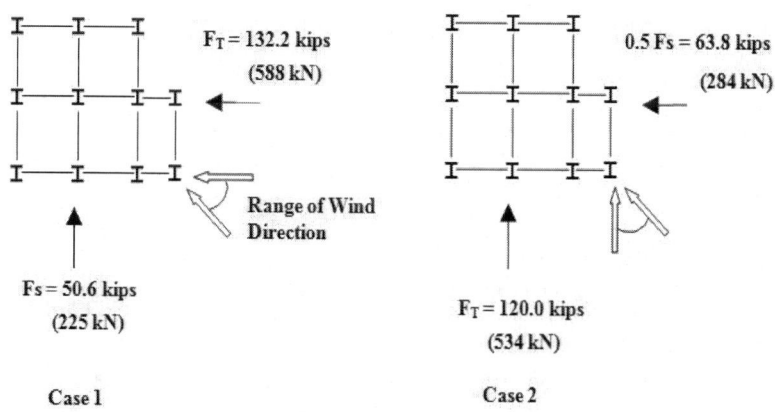

Wind in Two Principal Directions

Figure 6.2.1.1 Design Load Cases for Open Frame Example (Using Method of Section 5.2)

6.2.2 Open Frame Example – Using Method of Appendix 5A

The example problem of Section 6.2 will be reworked using the method of Appendix 5A. Refer to Section 6.2.1 as necessary for the previously calculated values.

6.2.2.1 Wind Toward Frame 3

$\varepsilon = 0.405$ $S_F / B = 0.488$ $N = 3$

From Equation 5A.1, $\alpha_{max} = (10 + 58\varepsilon)° = 33.5°$

From Figure 5A.3, for $\alpha_{max} = 33.5°$, $N = 3$, & $S_F / B = 0.488$ –
  For $\varepsilon = 0.1$: $C_f = 3.52$
  For $\varepsilon = 0.5$: $C_f = 2.53$

Interpolating for $\varepsilon = 0.405$ –
  $C_f = 3.52 - [(0.405 - 0.1) / (0.5 - 0.1) \times (3.52 - 2.53)] = 2.77$

Referring to Table 6.2.1.2: $F_s = (2.77 / 2.90) \times 127.6$ kips $= 121.9$ kips
Referring to Table 6.2.1.6: $F_E = 4.6$ kips
$\therefore F_T = 121.9$ kips $+ 4.6$ kips $= 126.5$ kips

Considering Perpendicular Direction –
  $90° - \alpha_{max} = 90° - 33.5° = 56.5°$

$\varepsilon = 0.235$ $S_F / B = 0.435$ $N = 3$

From Figure 5A.3, for $\alpha = 56.5°$, $N = 3$, & $S_F / B = 0.435$ –
  For $\varepsilon = 0.1$: $C_f = 2.25$
  For $\varepsilon = 0.5$: $C_f = 1.76$

Interpolating for $\varepsilon = 0.235$ –
  $C_f = 2.25 - [(0.235 - 0.1) / (0.5 - 0.1) \times (2.25 - 1.76)] = 2.08$

Referring to Table 6.2.1.8: $F_S = (2.08 / 3.44) \times 101.1$ kips $= 61.1$ kips

6.2.2.2 Wind Toward Frame A

$\varepsilon = 0.235$ $S_F / B = 0.435$ $N = 3$
From Equation 5A.1, $\alpha_{max} = (10 + 58\varepsilon)° = 23.6°$

From Figure 5A.3, for $\alpha_{max} = 23.6°$, $N = 3$, & $S_F / B = 0.435$ –
  For $\varepsilon = 0.1$: $C_f = 3.80$
  For $\varepsilon = 0.5$: $C_f = 2.34$

WIND LOADS FOR PETROCHEMICAL AND OTHER INDUSTRIAL FACILITIES 135

Interpolating for $\varepsilon = 0.235$ –
$C_f = 3.80 - [(0.235 - 0.1) / (0.5 - 0.1) \times (3.80 - 2.34)] = 3.31$

Referring to Table 6.2.1.8: $F_s = (3.31 / 3.44) \times 101.1$ kips $= 97.3$ kips
Referring to Tables 6.2.1.5 & 6.2.1.9: $F_E = 18.9$ kips
$\therefore F_T = 97.3$ kips $+ 18.9$ kips $= 116.2$ kips

Considering Perpendicular Direction –
$90° - \alpha_{max} = 90° - 23.6° = 66.4°$

$\varepsilon = 0.405$ $S_F/B = 0.488$ $N = 3$

From Figure 5A.3, for $\alpha = 66.4°$, $N = 3$, & $S_F / B = 0.488$ –
For $\varepsilon = 0.1$: $C_f = 1.42$
For $\varepsilon = 0.5$: $C_f = 1.27$

Interpolating for $\varepsilon = 0.405$ –
$C_f = 1.42 - [(0.405 - 0.1) / (0.5 - 0.1) \times (1.42 - 1.27)] = 1.31$

Referring to Table 6.2.1.2: $F_S = (1.31 / 2.90) \times 127.6$ kips $= 57.6$ kips

6.2.2.3 Open Frame Example (Using Method of Appendix 5A) -- Conclusion

The results of this example problem are for the frames only as shown in Figure 6.2.2.1. Calculation of loads on equipment is identical to method used in previous example.

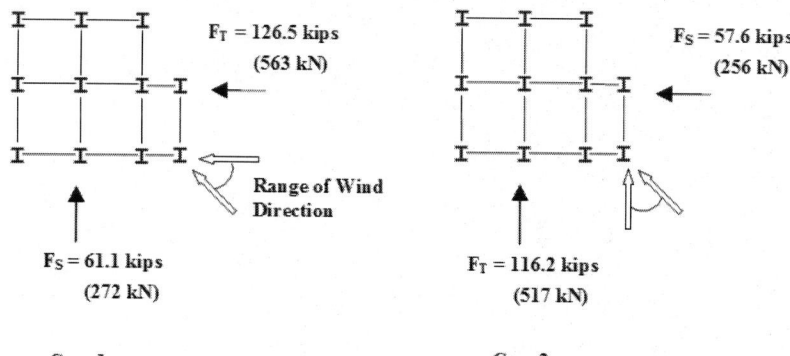

*Figure 6.2.2.1 Design Load Cases for Open Frame Example
(Using Method of Appendix 5A)*

6.2.3 Open Frame Example – Using Method of Appendix 5B

The example problem of Section 6.2 will be reworked using the method of Appendix 5B, assuming the structure to be a high-solidity open frame structure. Refer to Section 6.2.1 as necessary for the previously calculated values.

6.2.3.1 Wind Toward Frame 3

$L = 1/2 \times [(40 \text{ ft} + 1 \text{ ft}) + (49 \text{ ft} + 1 \text{ ft})] = 45.5 \text{ ft} \approx 46 \text{ ft}$ (average)
$B = 41 \text{ ft}$
$A_g = 3403 \text{ ft}^2$

$L/B = 46 / 41 = 1.12 < 1.5$

$\therefore C_f = 1/4 \times (L/B)^2 + 1.4$
$= 1/4 \times (1.12)^2 + 1.4 = 1.71$

The total wind force on the structure, F_T, will be calculated by application of Equation 5.1a, as shown in Table 6.2.3.1. Note that the calculated force, F_T, includes the wind load on the structure and appurtenances, equipment, piping, etc.

TABLE 6.2.3.1 Total Force - Structural Frame and Appurtenances, Equipment, Piping, etc. - F_T (Wind toward Frame 3)

Floor Level	Tributary Height (ft)	Width (ft)	A_g (ft)^2	q_z (psf)	G	C_f	F (lbs)
0	0 - 10 = 10	41	410	30.6	0.85	1.71	18,236
1	10 - 34 = 24	41	984	36.0	0.85	1.71	51,489
2	34 - 65 = 31	41	1271	41.4	0.85	1.71	76,482
3	65 - 83 = 18	41	738	44.0	0.85	1.71	47,198
						$F_T = \sum F =$	193,405

Note: To convert pounds force (lbs) to newtons (N), multiply F values in this table by 4.448.

6.2.3.2 Wind Toward Frame A

$L = 41 \text{ ft}$ $B = 46 \text{ ft}$ $A_g = 3844 \text{ ft}^2$

$L/B = 41 / 46 = 0.89 < 1.5$

$\therefore C_f = 1/4 \times (L/B)^2 + 1.4$
$= 1/4 \times (0.89)^2 + 1.4 = 1.60$

The total wind force on the structure, F_T, in this direction will be calculated similarly to that in 6.2.3.1 and is shown in Table 6.2.3.2.

WIND LOADS FOR PETROCHEMICAL AND OTHER INDUSTRIAL FACILITIES 137

Table 6.2.3.2 Total Force - Structural Frame and Appurtenances, Equipment, Piping, etc. - F_T (Wind toward Frame A)

Floor Level	Tributary Height (ft)	Width (ft)	A_g (ft)^2	q_z (psf)	G	C_f	F (lbs)
0	0 - 10 = 10	50	500	30.6	0.85	1.60	20,808
1	10 - 34 = 24	50	1200	36.0	0.85	1.60	58,752
2a	34 - 49 = 15	50	750	41.4	0.85	1.60	42,228
2b	49 - 65 = 16	41	656	41.4	0.85	1.60	36,935
3	65 - 83 = 18	41	738	44.0	0.85	1.60	44,162
						$F_T = \sum F =$	202,885

Note: To convert pounds force (lbs) to newtons (N), multiply F values in this table by 4.448.

6.2.3.3 Open Frame Example (Using Method Of Appendix 5B) -- Conclusion

The results of this example problem are shown in Figure 6.2.3.1. Note that in the load cases F_S is replaced by F_T for the orthogonal wind.

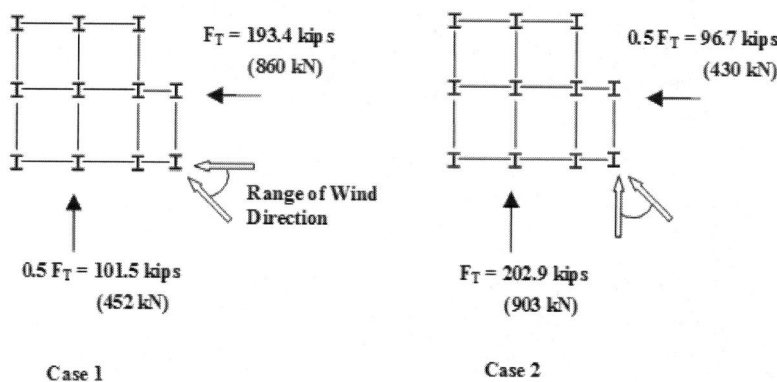

Figure 6.2.3.1 Design Load Cases for Open Frame Example (Using Method of Appendix 5B)

6.3 Partially Clad Structure Example

This section demonstrates the calculation of wind loads for partially clad structures per Section 5.3 of this guide. For this example, it is assumed that the stair tower for the structure depicted in Section 6.2 Figure 6.2.1, Figure 6.2.2, and Figure 6.2.3 is to be partially clad. Wind loads on the balance of the structure are not considered in this example, including any shielding effects, and therefore need to be calculated separately. Assume that the 9 ft. wide face along column line A and beyond column line 3 is to be clad from elevation zero to elevation 48 ft. (the full height of the stair tower). Also assume that the 20 ft. wide exterior face between column lines A and B

138 WIND LOADS FOR PETROCHEMICAL AND OTHER INDUSTRIAL FACILITIES

is to be clad for the full height. For the purpose of this example, assume that north is up on the diagrams for the open frame structure in Section 6.2. In this example the structure was classified as Category III and therefore I = 1.15.

Design wind forces are determined by Equation 5.1:

$F = q_z G C_f A_e$

The velocity pressure, q_z, is determined by *ASCE 7* Equation 6-15 for the mid height of the stair tower:

$K_z = 0.93$ for h = 24 ft (7.32 m)
$q_z = 0.00256 K_z K_{zt} K_d V^2 I = (0.00256)(0.93)(1.0)(0.85)(120)^2(1.15)$
    $= 33.51$ psf (1.60 kN / m^2)

The Gust Effect Factor, G, is equal to 0.85 per *ASCE 7* 6.5.8 for rigid structures.

Force Coefficients for a structure with two adjacent walls clad:

For wind directions for which the unclad faces are windward (wind from the north and west), $C_f = 2.0$ for these two directions simultaneously.

For wind directions for which the clad faces are windward (wind from the south and east), $C_f = 1.5$ for these two directions simultaneously.

The projected areas are the height times the width of each face.

For the face along column line A (wind from the north or south):
$A_e = 48$ ft x 9 ft = 432 ft^2 (40.1 m^2)

For the face located between column line A and B (wind from the east or west):
$A_e = 48$ ft x 20 ft = 960 ft^2 (89.1 m^2)

Wind force for each wind direction:

North Wind: F = 33.51 psf x 0.85 x 2.0 x 432 ft^2 = 24,610 lbs (110 kN)
East Wind: F = 33.51 psf x 0.85 x 1.5 x 960 ft^2 = 41,016 lbs (182 kN)
South Wind: F = 33.51 psf x 0.85 x 1.5 x 432 ft^2 = 18,457 lbs (82.1 kN)
West Wind: F = 33.51 psf x 0.85 x 2.0 x 960 ft^2 = 54,688 lbs (243 kN)

The application of the forces is shown in Figure 6.3.1. Two load cases are required to reflect the simultaneous application of North/West winds and South/East winds.

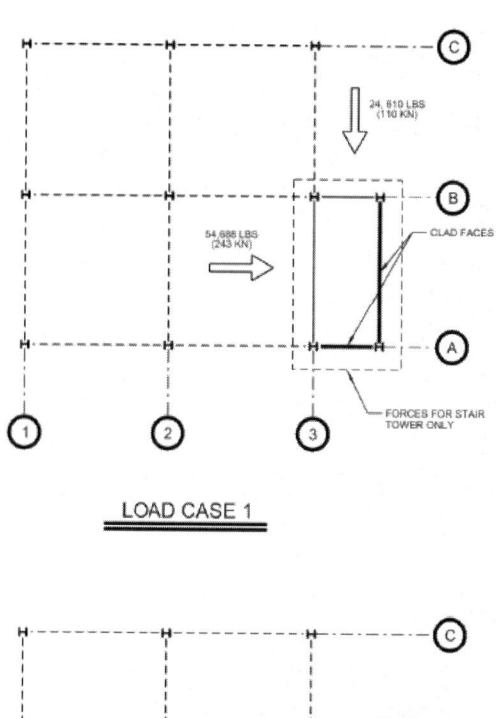
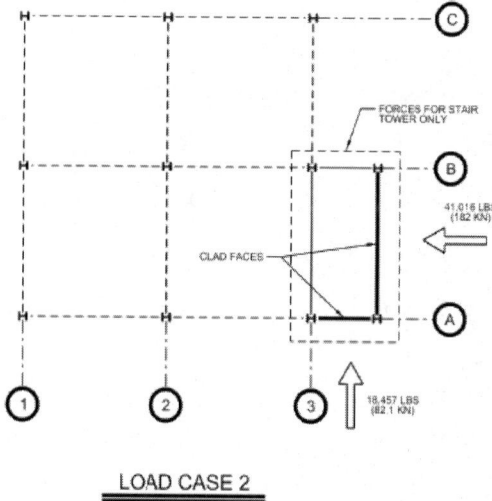

Figure 6.3.1 Partially Clad Structure Load Cases

6.4 Pressure Vessels Example

This section will demonstrate the application of the recommended guidelines for calculating wind loads on pressure vessels. In this example the vessel was classified as Category III and therefore an Importance Factor of I = 1.15 is required.

6.4.1 Vertical Vessel

Reference Figure 6.4.1 for description of vessel used in this example. Assume closest vertical vessel is spaced greater than three (3) times the vessel diameter. (Reference 5.4.1.4)

6.4.1.1 Simplified Method - Rigid Vessel

Wind loads for rigid vessels using simplified method determined by using Equation 5.1 in Section 5.0 (F = $q_z G C_f A_e$) are shown in Table 6.4.1. The velocity pressure, q_z is determined by 0.00256 x $K_z K_{zt} K_d V^2 I$ where K_d = 0.95. Other terms in this equation are determined as follows:

G = 0.85 (*ASCE 7* Section 6.5.8.1)

h/D = 150/10 = 15

Vessel is rough per Section 5.4.1.2, therefore

C_f = 0.8 + 8 x 0.1 / 18 = 0.84   (*ASCE 7* Figure 6-21)

Increased diameter to approximate appurtenances: (Section 5.4.1.2)

D + 5 ft. = 10 + 5 = 15 ft. or
D + 3 ft. + dia. of largest pipe = 10 + 3 + 1.5 = 14.5 ft.
Largest value controls, therefore, effective D = 15 ft. (4.57 m)

Therefore, A_e = 15Δh

Height increase to account for platform and vapor line above tangent line is one diameter, which is 10 ft. per section 5.4.1.2. Therefore, total effective height of the structure is 160 ft. (48.77 m).

Table 6.4.1 Simplified Method – Calculation of Base Shear

Ht. Above Ground	K_Z	q_z(psf)	G	C_f	A_e (ft.^{2})	F (lbs.)
0 - 15	0.85	34.2	0.85	0.84	225	5495
15 - 20	0.90	36.3	0.85	0.84	75	1944
20- 40	1.04	41.9	0.85	0.84	300	8975
40 - 60	1.13	45.5	0.85	0.84	300	9747
60 – 80	1.21	48.7	0.85	0.84	300	10432
80 -100	1.26	50.7	0.85	0.84	300	10860
100 -120	1.31	52.8	0.85	0.84	300	11310
120 -140	1.36	54.8	0.85	0.84	300	11738
140 – 160	1.39	56.0	0.85	0.84	300	11995

Total = 82496 lbs. (367 KN)

6.4.1.2 Detailed Method – Rigid Vessel

Vessel + Miscellaneous

Wind loads for a rigid vessel for the detailed method determined by using Equation 5.1 in Section 5.0 (F = $q_z C G_f A_e$) are shown in Table 6.4.2. The velocity pressure, q_z is determined by 0.00256 x $K_z K_{zt} K_d V^2 I$ where K_d = 0.95. Other terms in the equation are determined as follows:

G = 0.85   (*ASCE 7* Section 6.5.8.1)

h/D = 150/10 = 15

Per Section 5.4, determined that vessel is moderately smooth, therefore

C_f = 0.6 + 8 x 0.1 / 18 = 0.64   (*ASCE 7* Figure 6-21)

Increased diameter to approximate ladder, nozzles & piping 8" or smaller:

D + 1.5 ft. = 10 + 1.5 ft. (3.51m)   (Section 5.4.1.3)

Therefore, A_e = 11.5Δh

Table 6.4.2 Detailed Method – Vessel & Miscellaneous – Calculation of Base Shear

Ht. Above Ground	K_Z	q_z (psf)	G	C_f	A_e (ft.^{2})	F (lbs.)
0 - 15	0.85	34.3	0.85	0.64	172.5	3219
15 - 20	0.90	36.2	0.85	0.64	57.5	1132
20- 40	1.04	41.9	0.85	0.64	230	5243
40 - 60	1.13	45.5	0.85	0.64	230	5693
60 - 80	1.21	48.7	0.85	0.64	230	6093
80 -100	1.26	50.7	0.85	0.64	230	6344
100 -120	1.31	52.8	0.85	0.64	230	6606
120 -140	1.36	54.8	0.85	0.64	230	6857
140 – 150	1.39	56.0	0.85	0.64	115	3503

Total = 44,690 lbs. (199 KN)

Large Diameter Pipe (> 8")

Wind loads for piping determined using Equation, $F = q_z G C_f A_e$ are shown in Table 6.4.3, the velocity pressure, q_z, is determined by $0.00256 \times K_z K_{zt} K_d V^2 I$ where $K_d = 0.95$. Other terms in equation are determined as follows:

G = 0.85 (*ASCE 7* Section 6.5.8.1)

C_f = 0.7 (Section 5.4.1.3)

Pipe dia. = 18" = 1.5 ft. (0.46m)

Therefore, $A_e = 1.5\Delta h$

A_e above El. 150 = 3.14 x 10/2 x 1.5 = 24 ft.^{2} (2.2 m^2)
{Note - the pipe starts at El. 15.0 (see Figure 6.4.1)}

Table 6.4.3 Detailed Method Pipe – Calculation of Base Shear

Ht. Above Ground	K_Z	q_z (psf)	G	C_f	A_e (ft.^{2})	F (lbs.)
15 – 20	0.90	36.2	0.85	0.70	7.5	162
20 – 40	1.04	41.9	0.85	0.70	30	748
40 – 60	1.13	45.5	0.85	0.70	30	812
60 - 80	1.21	48.7	0.85	0.70	30	870
80 – 100	1.26	50.7	0.85	0.70	30	905
100 – 120	1.31	52.8	0.85	0.70	30	942
120 - 140	1.36	54.8	0.85	0.70	30	978
140 - 150	1.39	56.0	0.85	0.70	15	500
150 – 155	1.39	56.0	0.85	0.70	24	799

Total = 6,716 lbs. (29.9 KN)

WIND LOADS FOR PETROCHEMICAL AND OTHER INDUSTRIAL FACILITIES

Platforms (Refer to Figure 6.4.1)

The platform at El. 150 (just above the top of the vessel) is a square platform 12 ft. x 12 ft. in plan with handrail around the perimeter. It is assumed that the platform structural framing will be 8-inches deep (8"/12"/' = 0.70 sq. ft. / lin. ft.). The critical case will be one in which the wind direction is diagonal to the square platform, therefore, the length of the platform will be multiplied by sqrt of 2 (i.e., 1.414).

Therefore, A_e (platform framing) = 0.7 x 12 x 1.414 = 11.88 ft.^{2}
A_e (front handrail) = 0.8 x 12 x 1.414 = 13.58 ft.^{2}
A_e (back handrail) = 0.8 x 12 x 1.414 = <u>13.58 ft.^{2}</u>   (Section 5.4.1.3)
                                              39.03 ft.^{2} (3.62m^2)

$q_z = 0.00256 \times K_z \times K_{zt} \times K_d \times V^2 \times I = 50.0$ psf

$K_d = 0.85$

$G = 0.85$ (*ASCE* 7 Section 6.5.8.1)

$C_f = 2.0$ (Section 5.4.1.3)

$F = q_z G C_f A_e = 50.0 \times 0.85 \times 2.0 \times 39.03 = 3318$ lbs. (14.8 kN)

The other platforms are circular and extend 3 ft. beyond the outside radius of the vessel. Therefore the radial distance (R) from the centerline of the vessel to the outside of the Platform is 5 + 3 = 8 ft. (2.44 m). The angle (60°, 90° and 180°) shown on Figure 6.4.1 is the angle subtended by the ends of the platform as measured at the centerline of the vessel. Therefore, the projected length of the platform is calculated by the equation:

L = 2RSin (subtended angle/2)

Platform at El. 100 ft. - subtended angle 60°

Projected length = 2 x 8 x Sin (60/2) = 8.0 ft.

Assume platform framing is 6-in. deep (0.5 sq. ft. / lin. ft.)

A_e (platform framing) = 0.5 x 8.0 = 4.0 ft.^{2}

Ae (handrail)   = 0.8 x 8.0 = <u>6.4 ft.^{2}</u>
                                10.4 ft.^{2} (0.97 m2)

$q_z = 0.00256 \times 1.26 \times 0.85 \times (120)^2 \times 1.15 = 45.4$ psf

$G = 0.85$ (*ASCE* 7 Section 6.5.8.1)

144   WIND LOADS FOR PETROCHEMICAL AND OTHER INDUSTRIAL FACILITIES

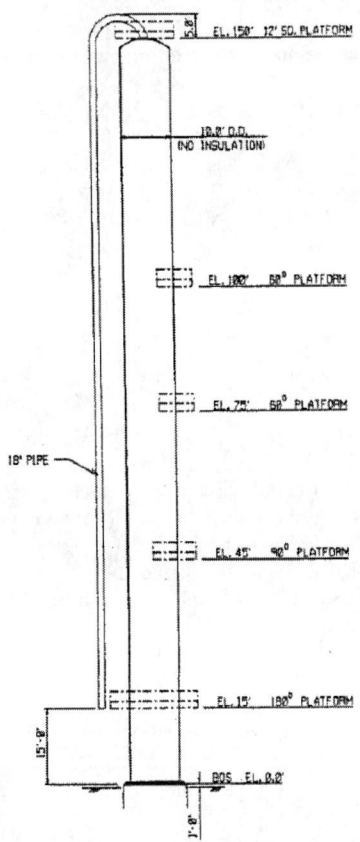

Shell thickness t = 1"
Empty Weight = 280 kips
Operating Weight = 500 kips

Figure 6.4.1 Vertical Vessel Example

$C_f = 2.0$ (Section 5.4.1.3)

$F = q_zGC_fA_e = 45.4 \times 0.85 \times 2.0 \times 10.4 = 803$ lbs. (3.6 kN)

Platform at El. 75 ft. – subtended angle = 60°

Projected length = $2 \times 8 \times \sin(60/2) = 8.0$ ft.

Assume platform framing is 6-in. deep (0.5 sq.ft. / lin.ft.)

A_e (platform framing) = $0.5 \times 8.0 = 4.0$ ft.^{2}
A_e (handrail) = $0.8 \times 8.0 = \underline{6.4 \text{ ft.}^2}$
$\phantom{A_e \text{(handrail)} = 0.8 \times 8.0 = }10.4$ ft.^{2} (0.97 m2)

$q_z = 0.00256 \times 1.21 \times 0.85 \times (120)^2 \times 1.15 = 43.6$ psf

$G = 0.85$ (*ASCE 7* Section 6.5.8.1)

$C_f = 2.0$ (Section 5.4.1.3)

$F = q_zGC_fA_e = 43.6 \times 0.85 \times 2.0 \times 10.4 = 770$ lbs. (3.4 kN)

Platform at El. 45 ft. - subtended angle = 90°

Projected length = $2 \times 8 \times \sin(90°/2) = 11.3$ ft. Since the projected length is larger than the vessel diameter, the back handrail will be included (Section 5.4.1.3). Back handrail project length = 3 ft. $\times \sin(45°) \times 2$ sides = 4.2 ft.

Assume Platform framing is 6-in, deep (0.5 sq.ft. / lin. ft.)

A_e (platform framing) = $0.5 \times 11.3 = 5.7$ ft.^{2}
A_e (front handrail) = $0.8 \times 11.3 = 9.0$ ft.^{2}
A_e (back handrail) = $0.8 \times 4.2 = \underline{3.3 \text{ ft.}^2}$ (Section 5.4.1.3)
$\phantom{A_e \text{(back handrail)} = 0.8 \times 4.2 = }18.0$ ft.^{2} (1.67 m2)

$q_z = 0.00256 \times 1.13 \times 0.85 \times (120)^2 \times 1.15 = 40.7$ psf

$G = 0.85$ (*ASCE 7* Section 6.5.8.1)

$C_f = 2.0$ (Section 5.4.1.3)

$F = q_zGC_fA_e = 40.7 \times 0.85 \times 2.0 \times 18.0 = 1245$ lbs. (5.5 kN)

Platform at El. 15 ft. - subtended angle = 180°

Projected length = 2 x 8 x Sin (180/2) = 16 ft. Since the projected length is larger than the vessel diameter, the back handrail will be included (Section 5.4.1.3). Back handrail project length = 3 ft. x Sin 90° x 2 sides = 6 ft.

Assume Platform framing is 6-in, deep (0.5 sq.ft. / lin. ft.)

A_e (platform framing) = 0.5 x 16.0 = 8.0 ft.^{2}
A_e (front handrail)   = 0.8 x 16.0 = 12.8 ft.^{2}
A_e (back handrail)   = 0.8 x 6.0 = 4.8 ft.^{2} (Section 5.4.1.3)
                                       25.6 ft.^{2} (2.38 m2)

q_z = 0.00256 x 0.85 x 0.85 x (120)^2 x 1.15 = 30.6 psf

G = 0.85   (*ASCE 7* Section 6.5.8.1)

C_f = 2.0   (Section 5.4.1.3)

F = $q_z G C_f A_e$ = 30.6 x 0.85 x 2.0 x 25.6 = 1332 lbs. (5.9 kN)

Total shear on platforms = 3318 + 803 + 770 + 1245 + 1332 = 7468 lbs. (33.22kN)

Total shear on foundation:

Vessel & Miscellaneous	=	44,684 lbs.
Large Diameter Pipe (> 8")	=	6,716 lbs.
Platforms	=	7,468 lbs.
Total	=	58,868 lbs. (261 kN)

6.4.1.3 Analysis of Flexible Vessels

The only difference between loads resulting from the analysis of the vessel as "Flexible" vs "Rigid" is that "G_f" (the Gust Effect Factor for main wind-force resisting systems of flexible buildings and structures) is substituted for "G" in the rigid analysis.

For the "Recommended" methods, the procedure outlined in *ASCE 7*, Section 6.5.8.2 was utilized to determine G_f = 1.099.

- SIMPLIFIED (FLEXIBLE):

    Total Shear = (G_f / G) x 82,493
                = (1.099/0.85) x 82,493 = 106,659 lbs. (475kN)

WIND LOADS FOR PETROCHEMICAL AND OTHER INDUSTRIAL FACILITIES 147

- DETAILED (FLEXIBLE):

    Vessel + Misc.:
    Shear = (G_f/G) x 44,684 lbs.
    = (1.099 / 0.85) x 44684 = 57,774 lbs.

    Pipe:
    Shear = (G_f/G) x 6716 lbs.
    = (1.099 / 0.85) x 6716 = 8,684 lbs.

    Platforms:
    Shear = (G_f/G) x 7468 lbs.
    = (1.099 / 0.85) x 7468 = 9,656 lbs.
    Total = 76,114 lbs. (339 kN)

Gust Effect Factor per *ASCE 7* (English Units)

Design Parameters:
    Wind Speed: V = 120 mph
    Importance Factor: I = 1.15
    Topographic Factor: K_{zt} = 1.0

Exposure Category: C   ASCE Table 6-2

$\bar{b}$ = 0.65   c = 0.20 ℓ = 500 $\bar{\epsilon} = \dfrac{1}{5}$ $\bar{\alpha} = \dfrac{1}{6.5}$

Z_g = 900   α = 9.5 Z_{min} = 15

Vessel Diameter: D = 10 ft. (excluding insulation)

Vessel Height: h = 150 ft.

First, consider the empty case:

Percentage Increase of Vessel Empty Weight: % increase = 1.1 x 280 kips = 28.0 kips (add 10% to account for piping and platform)

Vessel Empty Weight: W = (280 kip + 28kip) / 150 ft. = 2053 lb./ft.

Shell Thickness: t = 1.0 in.

Damping Ratio: β = 0.01

Vessel Period: $T = \dfrac{7.78}{10^6} \left(\dfrac{H}{D}\right)^2 \sqrt{\dfrac{12WD}{t}}$ T = 0.869 sec

Vessel Frequency h_1 = 1/T $h_1 = 1.151 s^{-1}$

148 WIND LOADS FOR PETROCHEMICAL AND OTHER INDUSTRIAL FACILITIES

The frequency is greater than one hertz and considered rigid for empty condition; therefore, use $G_f = 0.85$ per ASCE 6.5.8.1.

Now consider the operating case:

Vessel Operating Weight: W = (500 kip + 28.0 kip) / 150 ft. = <u>3520 lb./ft.</u>

Shell thickness: t = 1.0 in.

Damping Ratio: $\beta = 0.01$

Vessel Period: $T = \dfrac{7.78}{10^6}\left(\dfrac{H}{D}\right)^2 \sqrt{\dfrac{12WD}{t}}$ $\qquad$ T = 1.138 sec

Vessel Frequency $h_1 = 1/T$ $\qquad\qquad\qquad\qquad$ $h_1 = 0.879 s^{-1}$

The frequency is less than one hertz, so the vessel is considered flexible for the operating condition. G_f must be calculated according to ASCE 6.5.8.2.

Gust Effect Factor: (ASCE 6.5.8)

For flexible structures:

$$G_f = 0.925\left(\dfrac{1+1.7I_{\bar{z}}\sqrt{g_Q^2 Q^2 + g_R^2 R^2}}{1+1.7g_v I_{\bar{z}}}\right) \qquad (ASCE\ 7\ Eq.\ 6\text{-}8)$$

g_Q and g_V shall be taken as 3.4

$$g_R = \sqrt{2\ln(3600\eta_1)} + \dfrac{0.577}{\sqrt{2\ln(3600\eta_1)}} \qquad g_R = 4.159 \qquad (ASCE\ 7\ Eq.\ 6\text{-}9)$$

R, the resonant response factor, is given by:

$$R = \sqrt{\dfrac{1}{\beta}R_n R_h R_B (0.53 + 0.47 R_L)} \qquad (ASCE\ 7\ Eq.\ 6\text{-}10)$$

$$R_n = \dfrac{7.47 N_1}{(1+10.3 N_1)^{5/3}} \qquad R_n = 0.058 \qquad (ASCE\ 7\ Eq.\ 6\text{-}11)$$

Where $N_1 = \dfrac{\eta_1 L_{\bar{z}}}{V_{\bar{z}}}$ $\quad N_1 = 4.02$ $\qquad\qquad\qquad$ (ASCE 7 Eq. 6-12)

And, $\overline{V}_{\overline{z}} = \overline{b}\left(\dfrac{\overline{Z}}{33}\right)^{\alpha} V\left(\dfrac{88}{60}\right)$ $\overline{V}_{\overline{z}} = 133.49$ (*ASCE* 7 Eq. 6-14)

$R_\ell = \dfrac{1}{\eta} - \dfrac{1}{2\eta^2}\left(1 - e^{-2\eta}\right)$ (*ASCE* 7 Eq. 6-13a)

Where: $R_\ell = R_h$ setting $\eta = \dfrac{4.6\eta_1 h}{V_{\overline{z}}} = 4.543$ $R_h = 0.196$

$R_\ell = R_B$ setting $\eta = \dfrac{4.6\eta_1 B}{V_{\overline{z}}} = 0.303$ $R_B = 0.825$

$R_\ell = R_L$ setting $\eta = \dfrac{15.4\eta_1 L}{V_{\overline{z}}} = 1.014$ $R_L = 0.564$

Substituting these values into ASCE Eq. 6-10, the response factor, $R = 0.864$.

The intensity of turbulence at height $\overline{Z}$:

$I_{\overline{z}} = c\left(\dfrac{33}{\overline{Z}}\right)^{1/6} = 0.169$ (*ASCE* 7 Eq. 6-5)

Where $\overline{Z} = 0.6h \geq Z_{min}$

The background response Q is given by:

$Q = \sqrt{\dfrac{1}{1 + 0.63\left(\dfrac{B+h}{L_{\overline{z}}}\right)^{0.63}}}$ $Q = 0.887$ (*ASCE* 7 Eq. 6-6)

Where: $L_{\overline{z}} = \ell\left(\dfrac{\overline{Z}}{33}\right)^{\overline{\epsilon}} = 611$ in which ℓ and $\overline{\epsilon}$ are listed in Table 6-2 of *ASCE* 7.

Now, substituting all the variables into *ASCE* 7 Eq. 6-8:

$G_f = 0.925\left(\dfrac{1 + (1.7 * 0.169)\sqrt{3.4^2 * 0.887^2 + 4.159^2 * 0.864^2}}{1 + 1.7 * 3.4 * 0.169}\right)$

$\underline{G_f = 1.099}$

6.4.2 Horizontal Vessels

For description of the horizontal vessel used in this example, reference Figure 6.4.2.

6.4.2.1 Transverse Wind (wind on the side of the vessel)

$F = q_z G C_f A_e$ (A_e is defined in Section 5.4.2.2)

h = 20 ft. at platform level, k_z = 0.90, K_d = 0.95, q_z = 0.00256 x .95 x .90 x 120^2 x 1.15 = 36.24 psf

Vessel + Miscellaneous

G = 0.85   (*ASCE 7* Section 6.5.8.1)

B/D = 50/12 = 4.2, assume vessel is moderately smooth, therefore

C_f = 0.5 + 3.2 x 0.1/6 = 0.55   (*ASCE 7* Figure 6-21)

Increased diameter to approximate ladder, nozzles & piping 8" or smaller:

D + 1.5 ft. = 12 + 1.5 = 13.5 ft. (4.1m)   (Section 5.4.2.2)

Therefore, A_e = 13.5 x 54 avg. = 729 ft.^{2} (67.7 m^2)

Calculate Shear at Base:

$F = q_z G C_f A_e$ = 36.24 x 0.85 x 0.55 x 729 = 12353 lbs. (54.95 kN)

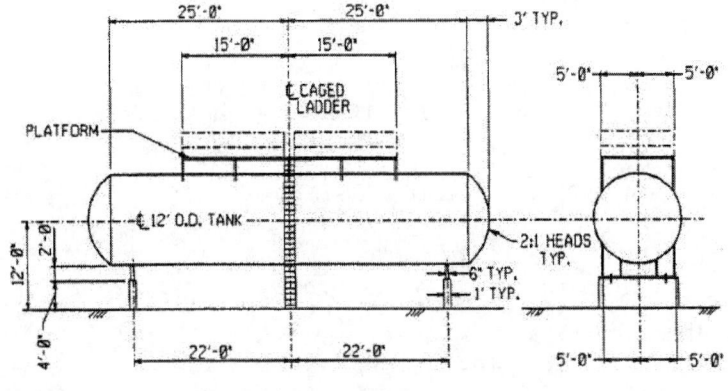

Figure 6.4.2 Horizontal Vessel Example

Platform

The platform is a rectangular platform 10 ft. x 30 ft. in plan with handrail around the perimeter. It is assumed that the platform structural framing will be 10-inches deep (10"/12"/' = 0.8 sq. ft. / lin. ft.).

Therefore, Ae (platform framing) = 0.8 x 30 = 24.0 ft.^{2}
 Ae (front handrail) = 0.8 x 30 = 24.0 ft.^{2}
 Ae (back handrail) = 0.8 x 30 = <u>24.0 ft.^{2}</u>   (Section 5.4.2.6)
 72.0 ft.^{2} (6.7 m^2)

q_z = 36.24 psf

G = 0.85 (*ASCE 7* Section 6.5.8.1)

C_f = 2.0 (Section 5.4.2.6)

F = $q_z G C_f A_e$ = 36.24 x 0.85 x 2.0 x 72.0 = 4440 lbs. (19.75 kN)

Supports

Steel Saddle:

A_e = 0.5 x 2.0 x 2 (supports) = 20.0 ft.^{2} (0.2m^2)

q_z = 36.24 psf

G = 0.85 (*ASCE 7* Section 6.5.8.1)

C_f = 2.0 (Section 5.4.2.6)

F = $q_z GCfA_e$ = 36.24 x 0.85 x 2.0 x 2.0 = 123 lbs. (0.55 kN)

Concrete support:

A_e = 1.0 x 4.0 x 2 (supports) = 8.0 ft.^{2} (0.7m^2)

q_z = 36.24 psf

G = 0.85 (*ASCE 7* Section 6.5.8.1)

C_f = 1.3 (Section 5.4.2.7)

F = $q_z G C_f A_e$ = 36.24 x 0.85 x 1.3 x 8.0 = 320 lbs. (1.5 kN)

152 WIND LOADS FOR PETROCHEMICAL AND OTHER INDUSTRIAL FACILITIES

Total Shear, Transverse Wind

Vessel & Miscellaneous	=	12,353 lbs.
Platform	=	4,440 lbs.
Supports (steel)	=	123 lbs.
Supports (concrete)	=	320 lbs.
Total	=	17,236 lbs. (76.67 kN)

6.4.2.2 Longitudinal Wind (wind on the end of vessel)

$F = q_z G C_f A_e$

$h = 20$ ft. at platform level, $q_z = 36.24$ psf

Vessel + Miscellaneous

$G = 0.85$ (*ASCE 7* Section 6.5.8.1)

Elliptical head, therefore $C_f = 0.5$ (Section 5.4.2.4)

Increased diameter to approximate ladder, nozzles & piping 8" or smaller:

$D + 1.5$ ft. $= 12 + 1.5 = 13.5$ ft. (4.1m)   (Section 5.4.2.2)

Therefore, $A_e = \pi \times 13.5 \times 13.5 / 4 = 143.1$ ft.^{2} (13.3m^2)

Calculate shear at base:

$F = q_z G C_f A_e = 36.24 \times 0.85 \times 0.5 \times 143.1 = 2204$ lbs. (9.8 kN)

Platform

The platform is a rectangular platform 10 ft. x 30. ft. in plan with handrail around the perimeter. It is assumed that the platform structural framing will be 10-inches deep (10"/12" = 0.8 sq. ft. / lin. ft.)

Therefore, Ae (plat. fram.)   = 0.8 x 10 =   8.0 ft.^{2}
             Ae (front handrail) = 0.8 x 10 =   8.0 ft.^{2} (Table 5.1)
             Ae (back handrail) = 0.8 x 10 =   <u>8.0 ft.^{2}</u>   (Section 5.4.2.6)
                                                     24.0 ft.^{2} (2.2 m^2)

$q_z = 36.24$ psf

$G = 0.85$ (*ASCE 7*, Section 6.5.8.1)

$C_f = 2.0$ (Section 5.4.2.6)

WIND LOADS FOR PETROCHEMICAL AND OTHER INDUSTRIAL FACILITIES

$F = q_z G C_f A_e = 36.24 \times 0.85 \times 2.0 \times 24.0 = 1479$ lbs. (6.6 kN)

Supports

Steel Saddle – 10 ft. wide x 3 ft. (avg.) high

$A_e = 10.0 \times 3.0 \times 2$ (supports) $= 60.0$ ft.^{2} (5.6m^2)

$q_z = 36.24$ psf

$G = 0.85$ (*ASCE 7*, Section 6.5.8.1)

$C_f = 2.0$ (Section 5.4.2.6)

$F = q_z G C_f A_e = 36.24 \times 0.85 \times 2.0 \times 60.0 = 3700$ lbs. (16.46 kN)

Concrete support – 11 ft. wide x 4 ft. high

$A_e = 11.0 \times 4.0 \times 2$ (supports) $= 88.0$ ft.^{2} (8.2m^2)

$q_z = 36.24$ psf

$G = 0.85$ (*ASCE 7*, Section 6.5.8.1)

$C_f = 1.3$ (Section 5.4.2.7)

$F = q_z G C_f A_e = 36.24 \times 0.85 \times 1.3 \times 88.0 = 3525$ lbs. (15.68 kN)

Total Shear, Longitudinal Wind

Vessel & Miscellaneous =	2,204 lbs.
Platforms =	1,479 lbs.
Supports (steel) =	3,700 lbs.
Supports (concrete) =	3,525 lbs.
Total =	10,908 lbs. (48.52kN)

6.5 Cooling Tower Example

Description: A wood frame, dual cell, cooling tower with metal siding and metal louvers is located at the industrial site near the Gulf Coast described in Section 6.0. The tower is 32 ft. long x 20 ft. wide x 12 ft. high. The fan cylinders are 6 ft. high and 8 ft. in diameter. Each metal siding panel is 2 ft. wide x 8 ft. long. The metal louvers are 3 ft. long and 1 ft. wide, with an inward slope of 45°. The cooling tower is shown in Figure 6.5.1. The fan deck flooring (i.e., roof) is 2 in. x 6 in. tongue and groove wooden planking material. The site is in a rural area. The tower lies along the edge of the plant, bordering a flat, open area that extends for half a mile. The basic wind speed per *ASCE 7* is 120 mph. The cooling tower services various administrative support facilities at the plant, and therefore the structure is Category II.

Figure 6.5.1. Cooling Tower for Example Problem

Using Method 1 – Simplified Procedure of *ASCE 7*: for a Mean Roof Height [15-ft. and assuming Exposure C, the Adjustment Factor λ will equal 1.21, with a Topographic Factor $K_{zt} = 1.0$, an Importance Factor I = 1.0, and a Roof Angle of 0°:

Main Wind Force Resisting System (MWFRS)

$p_s = \lambda \, K_{zt} \, I \, p_{s30}$ (*ASCE 7* Eq. 6.1 where p_{s30} is from *ASCE 7* Figure 6-2)

Horizontal Load on Walls:

Zone A: 10% x 20 ft. = 2 ft. < 0.4 x 12 ft. = 4.8 ft. (1.46 m)
2 ft. < 3 ft. min., so a = 3 ft. (0.91 m)

WIND LOADS FOR PETROCHEMICAL AND OTHER INDUSTRIAL FACILITIES   155

                Use: 2a = 2 x 3 ft. = 6 ft. Edge Zone from corners along both sides
                (1.21) (1.0) (22.8 psf) = 27.6 psf (1.35 kN/m^2)

Zone C: 32 ft. – 2 (6 ft.) = 20 ft. Interior Zone on long side (6.1 m)
              20 ft. – 2 (6 ft.) = 8 ft. Interior Zone on short side (2.44 m)
              (1.21) (1.0) (15.1 psf) = 18.3 psf (0.9 kN/m^2)

Vertical Load on Roof:

Zone E: Use: 2a = 6 ft. Edge Zone around perimeter of roof (1.836 m)
              (1.21) (1.0) (-27.4 psf) = -33.2 psf (uplift) (1.63 kN/m^2)

Zone G: 32 ft. – 2 (6 ft.) = 20 ft. Interior Zone on long side (6.1 m)
              20 ft. – 2 (6 ft.) = 8 ft. Interior Zone on short side (2.43 m)
              (1.21) (1.0) (-19.1 psf) = -23.1 psf (uplift) (1.13 kN/m^2)
              Note: 20 ft. x 8 ft. interior zone (Zone G) will coincide with fan openings.

Comment: The Simplified Procedure of *ASCE 7* designates two edge zones – Zone E and Zone F, with Zone E being the greater value; the procedure also designates two interior zones – Zone G and Zone H, with Zone G being the greater value; to simplify analysis of design wind loads on the MWFRS, the higher Zone E value is used for the edge zone, while the higher Zone G value is used for the interior zone. In most cases, the Zone G area will coincide with the fan openings.

Components and Cladding (C&C)

$p_{net} = \lambda\ K_{zt}\ I\ p_{net30}$ (*ASCE 7* Eq. 6.2 where p_{net30} is from *ASCE 7* Figure 6-3)

Metal Cladding Panel:

Effective Wind Area: 2 ft. x 8 ft. = 16 sq. ft. > 10 sq. ft., use 16 sq. ft. (1.49 m^2)

Zone 4: 10% x 20 ft. =   2 ft. < 0.4 x 12 ft. = 4.8 ft. (1.46 m)
    (Interior of Wall)   2 ft. < 3 ft. min., so a = 3 ft. (0.91 m)
                       20 ft. – 2 (3 ft.) = 14 ft. Interior Zone on short side
                                                             (4.27 m)
                        (1.21) (1.0) (25.2 psf) = 30.5 psf (inward)
                                                             (1.49 kN/m^2)
                        (1.21) (1.0) (-27.4 psf) = -33.2 psf (outward)
                                                              (1.62 kN/m^2)

Zone 5: a = 3 ft. End Zone from corners along short side (0.91 m)
(End of Wall)   (1.21) (1.0) (25.2 psf) = 30.5 psf (inward)
$$(1.49 \text{ kN/m}^2)$$
(1.21) (1.0) (-33.3 psf) = -40.3 psf (outward)
$$(1.97 \text{ kN/m}^2)$$

Metal Louver:

Effective Wind Area: 3 ft. x 1 ft. = 3.0 sq. ft. < 10 sq. ft., use 10 sq. ft. (0.93 m^2)

Zone 4: 10% x 20 ft. =   2 ft. < 0.4 x 12 ft. = 4.8 ft. (1.46 m)
(Interior of Wall)   2 ft. < 3 ft. min., so a = 3 ft. (0.91 m)
32 ft. − 2 (3 ft.) = 26 ft. Interior Zone on long side
$$(7.92 \text{ m})$$
(1.21) (1.0) (25.9 psf) = 31.3 psf (inward)
$$(1.53 \text{ kN/m}^2)$$
(1.21) (1.0) (-28.1 psf) = -34.0 psf (outward)
$$(1.67 \text{ kN/m}^2)$$

Zone 5: a = 3 ft. End Zone from corners along long side (0.91 m)
(End of Wall)   (1.21) (1.0) (25.9 psf) = 31.3 psf (inward)
$$(1.53 \text{ kN/m}^2)$$
(1.21) (1.0) (-34.7 psf) = -42.0 psf (outward)
$$(2.06 \text{ kN/m}^2)$$

Comment: Although the metal louver is inwardly inclined, air flow between the louvers is restricted, which could cause an increase in local pressure normal to the louver surface; moreover, separation of air flow along the leading edge of a louver may redirect and accelerate air flow toward an adjacent louver; hence, the total surface area of the louver is used without adjusting for the angle of inclination.

Uplift Load on Deck Planks:

Effective Wind Area: 10 ft. x 6"/12"/' = 5.0 sq. ft. < 10 sq. ft., use 10 sq. ft. (0.93 m^2)

Zone 1: 10% x 20 ft. = 2 ft. < 0.4 x 12 ft. = 4.8 ft. (1.46 m)
(Interior of Roof Deck)   2 ft. < 3 ft. min., so a = 3 ft. (0.91 m)
32 ft. − 2 (3 ft.) = 26 ft. Interior Zone on long side
$$(7.92 \text{ m})$$
20 ft. − 2 (3 ft.) = 14 ft. Interior Zone on short side
$$(4.27 \text{ m})$$
(1.21) (1.0) (10.5 psf) = 12.7 psf (inward)
$$(0.62 \text{ kN/m}^2)$$
(1.21) (1.0) (-25.9 psf) = -31.3 psf (outward)
$$(1.53 \text{ kN/m}^2)$$

WIND LOADS FOR PETROCHEMICAL AND OTHER INDUSTRIAL FACILITIES 157

Zone 2: a = 3 ft. End Zone on long and short sides (0.91 m)
(Ends of Roof Deck)   (1.21) (1.0) (10.5 psf) = 12.7 psf (inward)
$$(0.62 \text{ kN/m}^2)$$
$$(1.21) (1.0) (-43.5 \text{ psf}) = -52.6 \text{ psf (outward)}$$
$$(2.58 \text{ kN/m}^2)$$

Zone 3: a = 3 ft. Square Zone at all four corners (0.91 m)
(Corners of Roof Deck)   (1.21) (1.0) (10.5 psf) = 12.7 psf (inward)
$$(0.62 \text{ kN/m}^2)$$
$$(1.21) (1.0) (-65.4 \text{ psf}) = -79.1 \text{ psf (outward)}$$
$$(3.88 \text{ kN/m}^2)$$

<u>Horizontal Load on Fan Cylinder:</u>

$$p_z = K_z K_{zt} (V/28)^2 \, I \qquad\qquad\text{(Eq. 5.9)}$$

Projected Area = (6 ft.) (8 ft.) = 48 sq. ft. (4.46 m^2)

K_z = 0.88 assuming Exposure C at height = 12 ft. + 6 ft. = 18 ft. (5.49 m)

p_z = (0.88) (1.0) (120/28)^2 (1.0) = 16.2 psf across projected area (0.79 kN/m^2)

<u>Horizontal Load on Handrail:</u>

Follow method as described in Section 5.2.

6.6 Air Cooler (Fin-Fan) Example

Assume the air cooler geometry is as follows (see Figure 6.6):

- Single air cooler. 12 ft. (3.6 m) wide and 24 ft. (7.2 m) long and 4 ft. (1.2 m) high.
- The bottom of the air coolers (fin fans) are assumed to be 9 ft. (2.7 m) above the maintenance platform that is 5 ft. (1.5 m) above the top pipe level (elevation 35 ft.) with 4 ft. clearance underneath.
- Due to the varying shapes, configurations, and attachments, a solid shape will be assumed that matches the maximum dimensions of the equipment.

Design wind force on the air cooler is determined by Equation 5.1 (repeated below) where

$$F = q_z G C_f A_e$$

Design wind pressure q_z = 0.00256 $K_z K_{zt} K_d V^2 I$ (lb. / ft.^{2})   (*ASCE 7*, Section 6.5.10)
$\qquad\qquad\qquad\qquad$ = 0.613 $K_z K_{zt} K_d V^2 I$ (N / m^2)

From Section 6.0, V = 120mph, I = 1.0, and Exposure Category C.
Center of air cooler at elevation 46 ft. (Kz = 1.06)

$$q_{46} = 33.2 \text{ psf} (1.59 \text{ kN} / \text{m}^2)$$

For the wind direction perpendicular (transverse) to the pipe rack, the projected area for each air cooler is:

$$A_e = 12 \text{ ft. wide x 4 ft. high} = 48 \text{ ft.}^2 (4.5 \text{ m}^2)$$

From Table 5.3   a / b = 24 ft. / 12 ft. = 2.0
                   c / b = 4 ft. / 12 ft. = 0.33
                   $C_f = 0.91$

The force on the air cooler is

$$F = 33.2 \text{ psf x } 0.85 \text{ x } 0.91 \text{ x } 48 \text{ ft.}^2 = 1233 \text{ lbs. (5.5 kN)}$$

For the wind direction parallel to the pipe rack, the projected area for each air cooler is:

$$A_e = 24 \text{ ft. wide x 4 ft. high} = 96 \text{ ft.}^2 (8.9 \text{ m}^2)$$

From Table 5.3   a / b = 12 ft. / 24 ft. = 0.5
                   c / b = 4 ft. / 24 ft. = 0.167
                   $C_f = 1.0$

The force on the air cooler

$$F = 33.2 \text{ psf x } 0.85 \text{ x } 1.0 \text{ x } 96 \text{ ft.}^2 = 2709 \text{ lbs. (12.1 kN)}$$

For wind loads at access platforms and pipe rack see open structure examples.

WIND LOADS FOR PETROCHEMICAL AND OTHER INDUSTRIAL FACILITIES 159

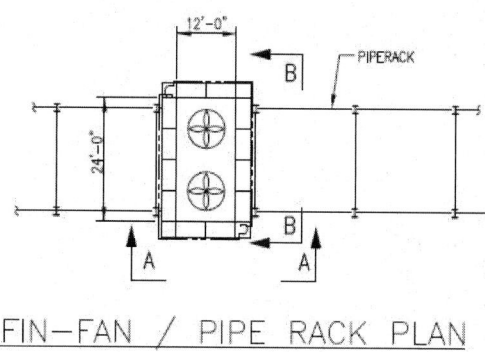

FIN—FAN / PIPE RACK PLAN

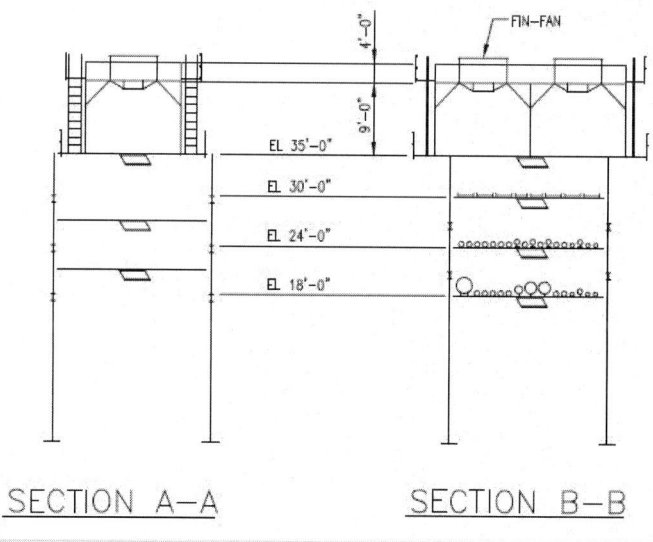

SECTION A—A   SECTION B—B

Figure 6.6 Air Cooler Example

REFERENCES

49 CFR 193.2067	Code of Federal Regulations, 49 CFR 193.2067, Transportation, Liquefied Natural Gas Facilities: Federal Safety Standards, Subpart B - Siting Requirements, Wind Forces.
Amoroso	Amoroso, S., Wind Loads for Petrochemical Structures, Ph.D. Dissertation, Louisiana State University, Baton Rouge, 263 p., 2007.
Amoroso and Levitan	Amoroso, S. and Levitan, M., "Recent Research for Wind Loads on Petrochemical Structures," Proceedings of the ASCE/SEI 2009 Structures Congress, Austin, TX, April, 2009a.
Amoroso and Levitan	Amoroso, S. and Levitan, M., "Wind Load Analysis Uncertainty for Petrochemical Structures," 11th Americas Conference on Wind Engineering, San Juan, Puerto Rico, June, 2009b.
Amoroso et al	Amoroso, S., Hebert, K, and Levitan, M., "Wind Tunnel Tests for Mean Wind Loads on Partially Clad Structures," Journal of Wind Engineering and Industrial Aerodynamics, doi:10.1016/j.jweia.2009.08.009, 2010.
Anderson	*Computational Fluid Dynamics, The Basics With Applications*, John David Anderson, McGraw-Hill, 1995.
ANSI A58.1	American National Standards Institute ANSI A58.1 *Minimum Design Loads for Buildings and Other Structures*, 1982.
ANSI/AWWA D 100-05	ANSI/AWWA D100-05, American Water Works Association, *Welded Carbon Steel Tanks for Water Storage*, May 1, 2006.
API 620	API 620, American Petroleum Institute, *Design and Construction of Large, Welded, Low-Pressure Storage Tanks, Eleventh Edition*, February 2008, Addendum 1, March 2009.

API 650	API 650, American Petroleum Institute, *Welded Steel Tanks for Oil Storage, Eleventh Edition*, June 2007, Effective date of May 2010.
AS/NZS 1170.2:2002	*Standards Australia/Standards New Zealand Structural Design Actions. Part 2: Wind actions.* AS/NZS 1170.2:2002. Standards Australia, Sydney, N.S.W., Australia, and Standards New Zealand, Wellington, New Zealand, 2002.
ASCE	*ASCE Wind Loads and Anchor Bolt Design for Petrochemical Facilities*, Task Committee on Wind Induced Forces and Task Committee on Anchor Bolt Design, American Society of Civil Engineers, New York, 1997.
ASCE 37-02	*Design Loads on Structures During Construction*, American Society of Civil Engineers, New York, 2002.
ASCE 67	*ASCE Wind Tunnel Studies of Buildings and Structures*, ASCE Manuals and Reports on Engineering Practice No. 67, American Society of Civil Engineers, New York, 1999.
ASCE 7	The 2005 edition of ASCE 7 and its predecessors. When noted by itself this reference is the 2005 version of this document. ASCE 7 is titled *Minimum Design Loads for Building and other Structures*.
ASCE Wind	*Wind Forces on Structures,* Transactions of the ASCE Vol. 126, Pages 1124-1198, 1962.
ASME	ASME STS-1–2006 (Revision of ASME STS-1–2000), *Steel Stacks*, American Society of Mechanical Engineers, New York, 2006.
Bailey and Levitan	James R. (Bob) Bailey and Marc L. Levitan, "Lessons Learned and Mitigation Options for Hurricanes," Process Safety Progress, Vol. 27, No.1, March 2008.
Calvert and Fouad	Calvert, Elizabeth, P.E. and H. Fouad, PhD., P.E., "A Review of Current Wind Load Provisions for Transmission Pole Structures," ASCE Structures, American Society of Civil Engineers, New York, NY, 2000.

Chen, et al	Chen et al, "Extended Boltzmann Kinetic Equation for Turbulent Flows," Science, Volume 301, 1 August 2003.
Chung	*Computational Fluid Dynamics*, T. J. Chung, Cambridge University Press, 2002.
CICIND	*Model Code for Steel Chimneys, Revision 1*, International Committee on Industrial Chimneys, Zurich, Switzerland, 1999.
Cook	Cook, N. J., *The Designer's Guide to Wind Loading of Building Structures Part 2: Static Structures*, Butterworths, London, 1990.
CTI	*CTI Code Tower Standard Specifications for the Design of Cooling Towers with Douglas Fir Lumber*, Cooling Technology Institute Bulletin STD-114, 1996.
Daugherty and Franzini	Daugherty, R. L. and Franzini, J. B., *Fluid Mechanics with Engineering Applications*, McGraw-Hill, 1977.
Dhamavaram, et al,	S. Dhamavaram, S. R. Hanna, and O. R. Hansen, "Consequence Analysis – Using a CFD Model for Industrial Sites," Process Safety Progress, Vol. 24, No. 4, December 2005, pp 316-327, 2005.
Ellingwood and Tekie	Ellingwood, B. and Tekie, P. B. "Wind load statistics for probability-based structural design." J. Struct. Eng., ASCE, 125(4), 453-463, 1999.
ESDU	ESDU. (1981) *Lattice Structures: Part 2: Mean Fluid Forces on Tower-like Space Frames, Data Item No. 81028*, Engineering Sciences Data Unit, London, 1981
Ferziger and Peric	Ferziger, Joel H., and Peric, Milovan, *Computational Models for Fluid Dynamics*, Springer-Verlag, Berlin, 2002.
Georgiou	Georgiou, P. N., "Wind Loads on Building Frames," M.E.Sc. Thesis, University of Western Ontario, *Canada*, 1979.

Georgiou, et al	Georgiou, P. N.; Vickery, B. J.; and Church, R. "Wind Loading on Open Framed Structures," Program and Workshop Notes, CWOWE III: Third Canadian Workshop on Wind Engineering, Vancouver VI.1 pp. 1-19, April, 1981.
Godoy	Luis A. Godoy, M. ASCE, "Performance of Storage Tanks in Oil Facilities Damaged by Hurricanes Katrina and Rita," ASCE Journal of Performance of Constructed Facilities, November/December 2007.
Hill and Levitan	Hill, Carol and Levitan, Marc, "Development of Vulnerability Functions for Industrial/Petrochemical Facilities due to Extreme Winds and Hurricanes," Coastal Disasters, American Society of Civil Engineers, New York, NY, 2005.
Humble	Corpus Christi Texas, on-site inspection performed by Humble Oil & Refinery Company, 1970.
IBC	International Building Code (IBC), 2009.
ICC 500/NSSA	*ICC/NSSA Standard for Design and Construction of Storm Shelters*, ICC-500, International Code Council, 2008.
J&H Marsh & McLennan	*Large Property Damage losses in the Hydrocarbon-Chemical Industries – A Thirty Year Review, Risk Control Strategies*, Eighteenth Edition, J&H Marsh & McLennan, 1998.
Kopp, et al	Kopp, G., Galsworthy, J., and Oh, J. "Horizontal Wind Loads on Open-Frame Low Rise Buildings," *Journal Of Structural Engineering,* American Society of Civil Engineers, v 136 (1), pp 98-105, 2010.
Levitan, et al	Levitan, M., Qiang, L., Amoroso, S., "Wind tunnel Tests on Open-Frame Industrial/Petrochemical Structures," Proceedings of the Fifth International Colloquium on Bluff Body Aerodynamics and Applications, Ottawa, July 2004.

Liu, et al	Journal of Wind Engineering and Industrial Aerodynamics, Volume 96, Issues 6-7, June-July 2008, Pages 831-839, Xianzhi Liu, Marc Levitan, Dimitris Nikitopoulos.
MBMA	MBMA, *Metal Buildings Systems Manual: 2010 Supplement to the 2006 Edition*, Metal Building Manufacturer's Association, Inc., Cleveland, OH.
Murakami and Mochida	Murakami, S., and Mochida, A., "Past, present, and future of CWE: the View from 1999," proceedings of the 10th International Conference on Wind Engineering, Copenhagen, Denmark, 21-24 June 1999.
Nadeem	Nadeem, A., "Wind Loads on Open Frame Structures", M.S. Thesis, Louisiana State University, Baton Rouge, Louisiana, 1995.
Nadeem and Levitan	Nadeem, A., and Levitan, M.L., "A Refined Method for Calculating Wind Loads on Open Frame Structures," Proceedings, Ninth International Conference on Wind Engineering (January 9-13; New Delhi, India), 1995.
Nadeem and Levitan	Nadeem, A, and Levitan, M. "A Refined Method for Calculating Wind Load Combinations on Open Frame Structures," Journal of Wind Engineering and Industrial Aerodynamics, 72, pp. 445-453, 1997.
NAS	*Hurricane Hugo: Puerto Rico, The U.S. Virgin Islands, and South Carolina, National Disaster Studies, Volume Six,* National Academy of Sciences, Washington, D.C., 1994.
NFPA 59A	NFPA 59A, *National Fire Protection Association, Standard for the Production, Storage and Handling of Liquefied Natural Gas (LNG).*
NIST 1476	NIST Technical Note 1476, *Performance of Physical Structures in Hurricane Katrina and Hurricane Rita: A Reconnaissance Report,* National Institute of Standards and Technology, Gaithersburg, MD 20899, June 2006.

Peterson	Petersen, R. L. "A wind tunnel evaluation of methods for estimating surface roughness length at industrial facilities'," Atmospheric Environment, v.31, No, 1, pp. 45 – 57, 1997.
PIP	Process Industry Practices (PIP), *Structural Design Criteria, STC01015 2007*.
Qiang	Qiang, L., "Wind Tunnel Tests for Wind Loads on Open Frame Petrochemical Structures," Master's Thesis, Louisiana State University, Baton Rouge, LA, 204 p, 1998.
Smith, et al	Smith et al, "Virtual Wind Tunnel Testing of Jack-up Legs," International Association of Drilling Contractors Conference, City University, London, 2001.
Strumolo and Babu	Strumolo, Gary, and Babu, Viswanathan, "New Directions in Computational Aerodynamics," Physics World, August 1997.
Walshe	Walshe, D. E., *"Measurements of Wind Force on a Model of a Power Station Boiler House at Various Stages of Erection,"* NPL Aero Report 1165, National Physical Laboratory, Aerodynamics Division, Teddington, UK, September, 1965.
Whitbread	Whitbread, R. E., *"The Influence of Shielding on the Wind Forces Experienced by Arrays of Lattice Frames"* Wind Engineering: Proceedings of the Fifth International Conference on Wind Engineering (Fort Collins, Colorado, USA, July, 1979), J. E. Cermak, Ed., Pergamon Press, 1980, pp. 405-420.
Willford and Allsop	Willford, M. R., and Allsop, A. C., *"Design Guide for Wind Loads on Unclad Framed Building Structures During Construction: Supplement 3 to The Designer's Guide to Wind Loading of Building Structures,"* Building Research Establishment Report, Garston, UK, 1990.

Index

aerodynamics: and cylinders 11; and open frame structures 9–10; and partially clad structures 10–11; and vertical vessels 11–14
air cooled heat exchangers 30, 95–96, 157–159
air coolers. see air cooled heat exchangers
ASCE 7 46–47
bridge structural members. see structural members, bridge
building codes: International Building Code 2006 60; International Building Code 2009 47–49, 60; International Existing Building Code 2006 60; International Existing Building Code 2009 61
cable trays: example calculations 113–114; force coefficients for 69; lateral loads on 119; longitudinal loads on 120–122; tributary area for 68
CFD models 45–46
computational fluid dynamics models. see CFD models
cooling towers 27–29, 94; example calculations 154–157; hurricane damage 37–38, 41–42
cross-wind forces 8–9
design practices 18–30; air cooled heat exchangers 30; cooling towers 27–29; existing practices 18–21; partially clad structures 22; pipe bridges 21–22; pipe racks 19, 21–22; steel stacks 25–27; tanks 22–25; uplift forces 23; wind loads on tanks 23
example calculations: air cooled heat exchangers 157–159; bridge structural members 117–118; cable trays 113–114; cooling towers 154–157; open frame structures 99–105, 123–137; partially clad structures 137–139; pipe bridges 117–122; pipe racks 111–117; pressure vessels 140–153; structural members 114–116; vertical vessels 140–149
fin-fans. see air cooled heat exchangers
force coefficients: for cable trays 69; for open frame structures 71, 72–77, 97–98, 106–107; for pipes 68; for structural members 68
horizontal torsion 82–83
hurricane damage: cooling towers 37–38, 41–42; power infrastructure 34–35, 40–41; refineries 32; tanks 32, 35–37, 38–39
hurricanes 31–43
IBC 2009. see International Building Code 2009
industry guidelines: ASCE 7 46–47; International Building Code 2009 47–49
International Building Code 2006 60
International Building Code 2009 47–49, 60
International Existing Building Code 2006 60
International Existing Building Code 2009 61
liquid natural gas facilities. see LNG facilities
LNG facilities 55–58
load combinations 46, 98, 107
loads, lateral 118–119
loads, longitudinal 119–122
loads, wind: codes, standards, and guides 14–15; formulation 3–6; research progress 15–17
petrochemical facilities 49–54
pipe bridges 21–22; example calculations 117–122; lateral loads on 118–119; longitudinal loads on

119–122
pipe racks 19, 21–22; example calculations 111–117; wind load analysis 67
pipes: force coefficients for 68; lateral loads on 119; longitudinal loads on 120–122; tributary area for 67
power infrastructure 34–35, 40–41
Process Industry Practices STC0105 Structural Design Criteria 59
Reynolds number 12
solidity 77–79, 106–109
spheres 93–94
steel stacks 25–27
structural members: example calculations 114–116; force coefficients for 68
structural members, bridge: example calculations 117–118; lateral loads on 118–119; longitudinal loads on 119–120
structures, open frame 69–83; aerodynamics of 9–10; along wind force calculations 127–130; alternative calculation methods 134–137; area of application of force 79–80; cross-wind force calculations 131–133; design load cases 80–82; example calculations 99–105, 123–137; force coefficients for 72–77, 97–98, 106–107; force coefficients for components 71; frame load 71–72; high-solidity 106–109; horizontal torsion 82–83; limitations of analytical procedures 72; load combinations 98, 107; main wind force resisting system 69–71; partially clad structures 83–85; solidity ratio 77–79
structures, partially clad 22; aerodynamics of 10–11; example calculations 137–139; wind load analysis 83–85

tanks 22–25; hurricane damage 32, 35–37, 38–39; wind loads on 23
tributary area: for cable trays 68; for piping 67
uplift forces 23
vessels, flexible 146–149
vessels, horizontal 92, 150–153
vessels, pressure 85, 140–153
vessels, rigid 140–146
vessels, vertical 86–88; aerodynamics of 11–14; appurtenances 13; aspect ratio 13; cylinder aerodynamics 11; directionality 14; example calculations 140–149; gust effects 14; neighboring vessels 13–14; Reynolds number 12; surface roughness 12
wind load analysis 31–65; CFD models 45–46; detailed method 90–91; evaluation on existing structures 58–64; historical performance of industrial structures 31–43; industry guidelines 46–49, 59–61; International Building Code 2006 60; International Building Code 2009 60; International Existing Building Code 2006 60; International Existing Building Code 2009 61; load combinations 46; past modifications 62–63; pipe racks 67; Process Industry Practices STC0105 Structural Design Criteria 59; simplified method 89; structural deterioration 63–64; uncertainty 64–65; wind exposure category 61–62; wind tunnel testing 43–44
wind loads: codes, standards, and guides 14–15; formulation 3–6; research progress 15–17
wind speeds 56–58; applicability of 57–58; determination of 56–57; interpretations 6–7